BIRMINGHAM UNIVERSITY LIBRARY

THIS BOOK MUST BE RETURNED
IMMEDIATELY IF RECALLED

TECHNOLOGY AND SOCIAL CHANGE
IN AMERICA

INTERPRETATIONS OF AMERICAN HISTORY

★ ★ ★   JOHN HIGHAM AND BRADFORD PERKINS, EDITORS

TECHNOLOGY AND SOCIAL CHANGE IN AMERICA

EDITED BY

Edwin T. Layton, Jr.

Case Western Reserve University

Harper & Row, Publishers

New York, Evanston, San Francisco, London

Sponsoring Editor: JOHN G. RYDEN
Project Editor: CHRISTOPHER J. KUPPIG
Designer: MICHEL CRAIG
Production Supervisor: VALERIE KLIMA

Excerpts from *Autobiography of an Idea* by Louis Sullivan reprinted by by permission of the American Institute of Architects.

TECHNOLOGY AND SOCIAL CHANGE IN AMERICA

Copyright © 1973 by Edwin T. Layton, Jr.
All rights reserved. Printed in the United States of America. No part of this book may be used or reproduced in any manner whatsoever without written permission except in the case of brief quotations embodied in critical articles and reviews. For information address Harper & Row, Publishers, Inc., 10 East 53rd Street, New York, N.Y. 10022.

Standard Book Number: 06-043881-9
Library of Congress Catalog Card Number: 73-8367

CONTENTS

Editor's Introduction vii

Introduction 1

Technology as Knowledge
 EUGENE S. FERGUSON 9

The Heroic Theory of Invention
 LOUIS C. HUNTER 25

The "American System" of Manufacturing
 ROBERT S. WOODBURY 47

The Direction of Technology
 BRUCE SINCLAIR 65

The Ideology of Technology
 HUGO A. MEIER 79

Technology and Government
 JOHN G. BURKE 99

Alienation and Technology
 LEO MARX 121

Science and Technology
 CARL W. CONDIT 131

Engineers in Revolt
  EDWIN T. LAYTON, JR.   147

Human Values and Modern Technology
  HERBERT J. MULLER   157

Selective Bibliography   175

Editors' Introduction

This volume—and companions in the series, "Interpretations of American History"—makes a special effort to cope with one of the basic dilemmas confronting every student of history. On the one hand, historical knowledge shares a characteristic common to all appraisals of human affairs. It is partial and selective. It picks out some features and facts of a situation while ignoring others that may be equally pertinent. The more selective an interpretation is, the more memorable and widely applicable it can be. On the other hand, history has to provide what nothing else does: a total estimate, a multifaceted synthesis, of man's experience in particular times and places. To study history, therefore, is to strive simultaneously for a clear, selective focus and for an integrated, over-all view.

In that spirit, each volume of the series aims to resolve the varied literature on a major topic or event into a meaningful whole. One interpretation, we believe, does not deserve as much of a student's attention as another simply because they are in conflict. Instead of contriving a balance between opposing views, or choosing polemical material simply to create an appearance of controversy, Edwin T. Layton, Jr., has exercised his own judgment on the relative importance of different aspects or interpretations of a problem. We have asked him to select some of what he considers the best, most persuasive writings bearing on technology and social change in America, indicating in the introductory essay and headnotes his reasons for considering these accounts convincing or significant. When appropriate, he has also brought out the relation between older and more recent approaches to the subject. The editor's own competence and experience in the field enable him to provide a sense of order and to indicate the evolution and complexity of interpretations. He is, then, like other editors in this series, an informed participant rather than a mere observer, a student sharing with other students the results of his own investigations of the literature on a crucial phase of American development.

JOHN HIGHAM
BRADFORD PERKINS

TECHNOLOGY AND SOCIAL CHANGE
IN AMERICA

Introduction

In 1900 an American historian, Henry Adams, visited the Paris Exhibition and found on display examples of the latest and most modern technology. Adams was particularly impressed by the dynamo, which he saw as a symbol of the modern age. He contrasted the dynamo with the Virgin, a symbol of an earlier time that had served to channel human energies in ways that served the highest beliefs of its age. But this was not true of the dynamo, which lacked any intrinsic human purpose. It produced power in its most abstract form and in ever-increasing quantities, but the uses of this power were undetermined. Adams associated this ever-increasing power with social disintegration.[1] The symbolism of the dynamo was apt: It expressed modern technology's lack of direction, its apparent neutrality or hostility to human values, and its disruptive impact on society. Like Adams, modern Americans are concerned with the negative effects of technology, and they are searching for ways and means of understanding and controlling this powerful force.

Adams was not typical of his fellow countrymen. Most Americans accepted and welcomed technological change with uncritical enthusiasm. Perhaps, no other nation in modern times has been so enamored of technology. The American's love affair with his automobile is proverbial.

The impact of technology in American history, however, has not been confined to the invention of conveniences and the elevation of folk heroes. At a time of widespread criticism of technology, it is well to bear in mind that technology has played vital roles in national development: in the conquest of a new continent, in the building of a great industrial nation, and in the achievement of the highest standards of living the world has ever

---

[1] Henry Adams, *The Education of Henry Adams, An Autobiography* (Boston and New York, 1918), 379–381. His later views of social disintegration, which he termed "dissolution" and "degradation," are in his essays "The Rule of Phase Applied to History" and "A Letter to American Teachers of History," in *The Degradation of the Democratic Dogma* (New York, 1919).

seen. Because we can take these things for granted, it is easy to stress technology's negative side. If Americans are quite properly concerned about the continued existence of poverty, ignorance, and disease in this country, is this not, at least in part, a reaction to the fact that technological development has made these things, for the first time, unnecessary? America's affection for technology rests on solid achievements.

Adams was not so much wrong as ahead of his time. Americans no longer accept technological change uncritically. As the adverse consequences of rapid, unrestrained technological change have become more and more apparent, a new mood of disillusionment and disenchantment has replaced the earlier enthusiasm. Technology is rightly associated with environmental pollution, the failures of our cities, and a thousand and one other factors that make our society a less pleasant and less safe place in which to live. Day-to-day frictions such as congestion, noise, and smog, are clearly the reflections of deeper problems: Our national priorities seem distorted. Television, one of the finest triumphs of technology, daily inculcates materialism, status seeking, and other false values. A large portion of our national wealth and ingenuity goes into the development of weapons of mass destruction. Clearly, something is wrong.

To some, technology itself is at fault. To them it has taken on the nightmarish qualities of a demonic force threatening to destroy society. Jacques Ellul, a French scholar, has expressed the views of many American and European intellectuals: Technology is a mysterious force that cannot be controlled. It is destroying, inexorably and inevitably, almost every shred of human dignity and every worthwhile human value. The end is a science fiction horror, a society in which men are ruled by machines.[2] This view is founded on ignorance. Technology is not some mysterious demiurge; it is a social process conducted and directed by men.

Though technology has been one of the most important forces influencing American society, the sad truth is that we know very little about it. Our fundamental misconception of technology is responsible for much of this ignorance. Historians have assumed that technology consists only of machines and gadgets. And thus, since history is the story of men, not hardware, the role of technology in history has been unobvious. Historians and social critics often treat technology as something lying outside history, with society as its passive object. This leads to determinism. In America, the pessimists like Adams and Ellul have been in the minority. A more

[2] Jacques Ellul, *The Technological Society* (New York, 1964).

popular approach, taken by writers like Roger Burlingame, has been to portray American development in terms of triumphant gadgeteering.³

In the last dozen years, a group of professional historians has been attracted to the history of technology, partially in response to contemporary disillusionment with technology. In their work we detect a subtle, but profound difference in emphasis: To them technology is not so much the story of machines as it is the story of knowledge. Technology is not, however, simply knowledge in the abstract. It is knowledge at work—the knowledge that gives men the ability to do things. Technology, therefore, is knowledge operating in a social context and finding expression in machines and tools. Technology is imbedded in work; its study involves humble folk—craftsmen, farmers, and working men generally—not just famous inventors, engineers, and businessmen. Indeed, technology need not involve machines at all—the creation and dissemination of useful knowledge is as much a part of technology as are tools. As knowledge at work, technology is a social phenomenon subject to historical inquiry. From a critical study of the social role of technology, we may hope to gain a deeper understanding of American history. A better understanding of how technology has operated may enable Americans to control technology and direct it more effectively for human benefit.

If technology is not a mysterious dictator, then what is its historical role? Broadly, there have been two answers to this question. Some writers have said that technology is the means for man's conquest of nature. Others, in reply, assert that technology is a tool that some men use to control other men. Both are right. As A. Hunter Dupree has argued, technology constitutes an "information system" that mediates between man and his natural environment.⁴ It is, therefore, a key element in social adaptation. But, this is not its only function; technology is part of a larger entity—culture—and as such it influences and is influenced by other aspects of society. Consequently, it is a fundamental element in social evolution. And some men do indeed use technological changes to gain control over other men. Examples of this dual function of technology are easy to find. The railroads enabled Americans to conquer distance and

---

³The standard survey of technology in America is still Burlingame's two volumes, *March of the Iron Men* (New York, 1938) and *Engines of Democracy* (New York, 1940).

⁴A. Hunter Dupree, "The Role of Technology in Society and the Need for Historical Perspective," *Technology and Culture*, 10 (October 1969), 528–532.

played an integral part in the transformation of America from a small, decentralized agrarian republic to a giant, centralized industrial and urban society. They also provided opportunities for immense profit to a few. By controlling them, a small group of men were able to make decisions that affected, sometimes adversely, the lives of many other Americans.

Corresponding to the dual nature of technology's relation to society, there are two branches of the history of technology: the *internalist* and the *externalist*. Internalists are concerned with the origins and nature of technological change. Externalists are concerned with the effects of technology on other parts of society. These two approaches are not mutually exclusive; indeed, they depend upon each other. Internalists must understand the social causes of technological change; they must examine the complex network of ideas, institutions, and policies that produce new technologies. Conversely, a clear understanding of the nature of particular technological changes is necessary before one can discover how these changes influence society.

There is a tendency to apply the internalist approach to earlier American history, roughly to the Civil War, while the externalist analysis is focused on more recent American history. This division is rather arbitrary and should disappear with time; but it is not without its own pragmatic logic. Early Americans were chiefly concerned with technology as a tool for the conquest and development of their country. The emphasis was upon encouraging technology to meet national needs. Since the Civil War, however, Americans have become increasingly concerned with the social consequences of technological change. On a purely practical level, recent technology presents internalists with difficulties because of its complexity and scientific sophistication. Externalists, on the other hand, are drawn to the modern period not only because of the magnitude of the changes and their social importance, but also because of the greater abundance of source materials for recent history. But the relation between technology and society is a two-way street, and both internalist and externalist approaches are necessary for a balanced account. As the number of historians of technology grows from the present handful, we may expect a broadening and deepening of historical understanding. So far we have only a very sketchy preliminary outline of the role of technology in American history.

Though the history of technology is still in an early stage of development, its importance is already clear. There has not been time for an overall synthesis, but a great deal of highly suggestive work is imbedded in the journal literature and in monographs, where, unfortunately, it is not readily available to students of American history. This anthology is an attempt to remedy this situation. It falls into two parts. The first five selections deal with the origin and direction of technological change; the second five examine the impact of technology on society. The division between externalists and internalists, however, is not sharp. The editor has tried to select articles that exhibit the complementary nature of these two approaches.

Most of the writings on the history of technology in America center on the introduction of modern technology—the process we call the Industrial Revolution. One of the important discoveries made by modern scholars is that the Industrial Revolution was not the same in all countries. Of course, some things were constant: The steam engine, the factory system, and the growth of cities were common to all countries. But the differences were important. America was more successful than most countries in using modern technology to meet its needs. One reason for this was that Americans were able to adapt major technologies and assimilate them within distinctive American traditions. The use of the steamboat on western rivers and the "American System" of manufacturing with interchangeable parts are striking examples, Both were quickly recognized as important American contributions to technology.

In explaining historical events, such as an important technological innovation, there is a tendency to give the credit to a hero. Or in the case of unpleasant events such as wars, we speak of a "devil theory of history." Therefore, one of the most important tasks of the first generation of critical scholars in any field of history is to debunk the myths and legends that encrust the records of the past. The study of American technological development is no exception. The heroic theory of invention attributes important innovations to a few seminal figures—the Whitneys and Fords of legend. This viewpoint continues to be a serious impediment to real understanding. The "flash of genius" is hard to analyze, and uncritical biographies tend to interpret inventors in terms of American mythology. Eugene Ferguson is typical of modern historians of technology. He sees technology as knowledge. His concern is not the isolated stroke of genius but patterns in the flow of information. His emphasis is not the isolated inventor, but an interacting community of technologists.

The study of technological change compels us to distinguish between invention, a mental act, and innovation, a socioeconomic process. Studies like those of Louis C. Hunter and Robert C. Woodbury suggest that many important innovations, such as the steamboat and the "American System" of manufactures, are not single inventions but complex combinations of many inventions. Such innovations evolve over time through the labors of many men, not just one or two heroic figures. Innovations arise out of social needs, and they are guided by social forces. In these terms, technological change becomes a process amenable to historical investigation.

Rejecting the heroic theory of invention has provided many historical dividends. Machines and tools, like books and works of art, tend to reflect the society that produces them. They can tell us a great deal about American life. The abandonment of the heroic theory does not eliminate the inventor from the story, any more than a modern account of the American Revolution would omit George Washington. The difference lies in the fact that major figures are no longer isolated. They can be seen as part of a larger whole, acting in a rich context of ideas, institutions, and men.

Instead of worshipping the heroes of legend, modern historians investigate the forces that guide and direct technological change. Technology, as knowledge, depends on institutions that accumulate and disseminate information. Bruce Sinclair has studied an important early example: the Franklin Institute. He examines the role of the Franklin Institute in fostering an American style in technology. In so doing, he examines the important social realities that lay behind the seemingly minor issue of American preferences in screw threads. Hugo A. Meier looks at another side of the story. Technological change depends on support from the community. Meier examines the attempted reconciliation of the new technology with American democracy. These efforts produced a set of powerful ideological supports for technological change that help account for the resistance to the introduction of modern technology in America.

Externalists are less concerned with the process of technological change than with the social effects produced by innovations. Older determinists like Roger Burlingame viewed history in terms of the theory of *social lag* developed by the sociologist William F. Ogburn. Technology produces changes, but society lags behind and has to catch up; the inevitable delays produce dislocations and problems.[5] Within this framework, it is often useful to speak of the *primary* and *secondary* effects of innovations.

[5]George H. Daniels, "The Big Questions in the History of American Technology," *Technology and Culture*, 11 (January 1970), 2–3.

Primary effects are usually planned or foreseeable, involving the fulfillment of the ends originally sought. Secondary effects, which economists call *negative externalities*, are the long-range, unplanned, and unforeseen changes produced by technological activity. Most of the adverse effects of technology fall into this category.

Stripped of its metaphysical barnacles, the theory of social lag and its separation of primary and secondary effects is of great value in analysis. John Burke's study of the problem of steam boiler explosions is a case in point. The primary effect of the development of the steamboat was a means of utilizing America's vast river system for cheap transportation. But these boats created a hazard to the public because of boiler explosions, leading ultimately to government regulation—a secondary effect not foreseen by the rugged individualists responsible for the innovation. In this case, the long-range effect was scarcely predictable at the outset, since the first steamboats used low-pressure engines that were not prone to explode. It was the subsequent introduction of high-pressure engines that was responsible for the problem. Nor was it obvious that the steamboat was only the first of a series of technological changes that would force government intervention into the economy. But railroads, gas and electric utilities, electronic communications, the automobile, the airplane, and a host of other new technologies have all added their increment of change and social dislocation. The result was the gradual collapse of laissez-faire policy in America and the adoption of positive regulatory measures.

Though the idea that technology has both primary and secondary effects is often useful, it does not tell the entire story. There are cases in which the secondary effects were totally unpredictable. But there are many more in which the results are in some sense implicit in the original situation. In a society based on free enterprise capitalism, technological change is guided mainly by private individuals acting for profit. In other words, social effects, other than narrowly economic ones, are systematically excluded from the process of technological decision making. Therefore, the adverse results of technology are not simply a matter of accident. Implicit assumptions and biases are built into the mechanisms of technological change. Alienation provides an example of an unforeseen result that was ironically implicit in the original situation. Leo Marx has studied this slippery subject. He traces the sources of disenchantment in the writings of Hawthorne and Melville to a separation of head and heart, which these authors found in the technology of their day. But this separation was no accident. It reflected a system that directed technology by rationality and profit and not by a concern for human values.

Besides the profit system, modern science has provided important guidance for technological development. Carl Condit examines a basic change in the technological community in the nineteenth century. Science and technology were all but merged, and technology took on most of the attributes of science. This enhanced the power of technology. It also accelerated the rate of technological change and of social dislocation. The new scientific technologists saw reality in terms of matter and motion, not man and his feelings. And the works designed by these men tended to reflect not so much human values as the abstract world of modern science, providing yet another source of modern alienation.

As the adverse effects of technology have become more and more apparent, the issue of the control of technology has come to the fore; it is now a major political issue. Among the first to see this problem was Thorstein Veblen, a great philosopher of technology, along with certain members of the engineering profession. But, as the paper by Edwin Layton indicates, the motives of each were different. Veblen wanted a radical redirection of society, while the engineers were concerned with marginal changes that would prevent radical change, while advancing their own professional interests. Herbert J. Muller's selection suggests some of the current perplexities produced by the failure of our society to find a consensus that would make technology serve human values.

Technology as Knowledge

EUGENE S. FERGUSON

Eugene S. Ferguson, professor of history of technology at the University of Delaware and the Hagley Museum, is one of the founding fathers of the contemporary discipline of the history of technology. His *Bibliography of the History of Technology* (Cambridge, Mass., 1968) is an indispensable guide for any serious student. The present article is based on his editing of the *Early Engineering Reminiscences, 1815–1840, of George Escol Sellers*, U.S. National Museum Bulletin no. 238 (Washington, D.C., 1965). In all these works, we see the professional historian at work, searching for original sources and asking new questions. In accounting for the remarkable advance made by American technology in the first half of the nineteenth century, Ferguson draws attention not to the isolated inventor, but to the existence of a mechanical community. The emphasis shifts from particular machines to the accumulation and dissemination of knowledge. Henri Morgeme's maxim is as relevant for the historian as it was for Sellers: "no understand without beginning right."

The term "know-how" appeared in the United States in the twentieth century. It was invented to describe the peculiar set of constructive and organizational abilities that are essential ingredients in the building of an industrial complex. Although the term is new, it is fairly evident that the abilities it describes are not. But when we ask for particulars of the origin and development of "know-how," the only answers that have been formulated are couched in economic terms. The objection that complex and

Eugene S. Ferguson, "On the Origin and Development of American Mechanical 'Know-How,'" *Midcontinent American Studies Journal*, 3 (Fall 1962), 3–15. Copyright © 1962 by the Midcontinent American Studies Association. Reprinted without footnotes by permission.

sophisticated tools and machines were required in order to build an economy of plenty is dismissed usually by a vague reference to Yankee ingenuity.

An examination of the rise of American technology revals the need for more plausible explanations than we have been accustomed to. It is the purpose of this article to suggest the kind of question that might be asked and to point to what appear to me to be fruitful lines of inquiry.

The tradition of native American mechanical "know-how" was well established by the middle of the nineteenth century. Citizens of the United States were acutely aware of the mechanical prowess of their "go-ahead" country. The nation was, indeed, going forward. In *Harper's Magazine*, the Easy Chair Editor touched the dominant chord when, in 1853, he wrote "Our fast age is growing rapidly faster." The American "brag" was commonly remarked on both sides of the Atlantic. At the beginning of the same century—just fifty years earlier—however, the pace of material advance can only be described as halting. The seeds of know-how were present, of course; they had been imported from western Europe; but they had only just begun to germinate.

The contrast between the mechanical capabilities of craftsmen in 1800 and in 1850 is so striking that it would appear to demand an explanation. Yet little attention has been paid to this fundamental development which underlay the spectacular material achievements of these years. I have treated the subject in a personal and informal way because nowhere has enough information been assembled to make any final pronouncements, and because I should like to suggest the large number of variables that are involved in the increase and diffusion of mechanical knowledge.

First, let us look at the nature of the contrast just mentioned. In 1800, the Philadelphia waterworks were under construction. Designed by Benjamin H. Latrobe, an English-trained architect and engineer, the works required two steam pumping engines. One engine, located on the banks of the Schuylkill River, pumped water from the river to the second engine, which was in the main waterworks building in the Center Square, where City Hall now stands. The engines were of the conventional Boulton and Watt design; the largest engine cylinder that had to be made was just under 40 inches in diameter and about $6\frac{1}{2}$ feet long. The contract for building the engines was let to Nicholas Roosevelt, of Newark, New Jersey, whose machine works—called Soho Works, after the Boulton and Watt shops in Birmingham, England—were as advanced as any in the

United States. The engines were completed; they operated satisfactorily for many years, but their construction taxed Roosevelt's shops to the limit. A visitor to his establishment during the summer of 1800 reported on the progress of boring one of the large engine cylinders.

The cylinder, of cast iron, had to have about ¾-inch of material removed in order to obtain a smooth and true interior cylindrical surface. The boring mill, arranged horizontally, consisted of a boring head, driven by a water wheel, and a movable carriage to which the cylinder was fastened and which served to advance the cylinder against the boring head.

> Two men are required [wrote the visitor]. One almost lives in the cylinder, with a hammer in hand to keep things in order, and attend to the steelings [the cutting tools], the other attends to the frame on which the cylinder rests which is removed by suitable machinery; these hands are relieved, and the work goes on day and night; one man is also employed to grind the steelings; the work is stopped at dinner time, but this is thought no disadvantage as to bore constantly the cylinder would become too much heated; the work also stands whilst the steelings are being changed, which requires about ten minutes time, and in ten minutes more work they were dull again. . . . The workmen state that the boring was commenced on the 9th of April and had been going on ever since, three months, and about six weeks more will be required to finish it.

In December, 1852, in New York City, the main propulsion engine of the caloric ship *Ericsson* had been installed on board, and last minute adjustments were being made for the trial run of the vessel, which occurred in the early days of 1853. The engine of the *Ericsson*, which was designed by the Swedish-American John Ericsson, was one that employed heated air, rather than steam, as the working medium. Ericsson expected that this "caloric" engine would make steam engines obsolete, and there were indeed some novel design features involved; but the most remarkable fact about the *Ericsson's* engine was its size.

Four cylinders, each 14 feet in diameter and perhaps 8 feet long (the working stroke was 6 feet), and four more cylinders, each 11½ feet in diameter, were components of the enormous caloric engine. The cylinders were cast and bored at the machine works of Hogg and Dalamater, which was one of several large mechanical firms located in lower Manhattan.

The details of this particular boring operation have not been discovered; but it is probable that the boring mill was arranged vertically and driven by a steam engine, that the cutting tools were greatly improved over those used fifty years earlier, and that the boring proceeded at a rate that would have amazed Roosevelt. A contemporary observer who was not inclined to be generous in his comments was Orson Munn, editor of Scientific American. He thought the "caloric" feature of the engine a humbug, but he had only praise for the mechanical design and machine work. Munn wrote that "the designer and constructors of [the] machinery have shown themselves to have long heads, and skilful hands. We have never seen anything to compare with the castings."

So far as I can learn, these were the largest engine cylinders ever attempted, anywhere, before or since. This kind of performance, involving construction on an heroic scale, was typical of the United States in the 1850s.

Returning now to 1800, we can look briefly at another kind of mechanical performance.

Eli Whitney, of New Haven, Connecticut, had engaged, on June 14, 1798, in a contract to supply 10,000 muskets to the federal government within a period of just over two years. He expected within the first year to build machinery that would enable him to use unskilled labor to produce muskets in quantity; but by the end of 1800 he had delivered no muskets, although his contract period had run out. The first 500 muskets were delivered late in 1801, and the contract was not closed until January, 1809. In no single year before 1806 did he deliver more than 1000 muskets.

Now this is the performance that is usually accepted uncritically as evidence of Whitney's role as the father of mass production industries. Upon closer examination it will be found that not only did other private gun contractors and the national armories at Springfield, Massachusetts, and at Harper's Ferry furnish many more muskets than did Whitney during this period, but that their methods were generally superior to those employed by Whitney. The idea of a system of progressive manufacture of interchangeable parts, using special-purpose machines to aid and, where possible, replace the skill of the operator, was present in 1800. The idea originated probably in France; it was known perhaps in England; but nowhere had it been effectively applied.

In 1853, on the other hand, the American inventor Samuel Colt was operating a pistol manufactory in London, using special-purpose machines that he had brought with him from the United States; and in 1854 a British commission, representing the Board of Ordnance, was in the United States to inquire into what the British called the "American

System" of small-arms production and to purchase machine tools from American manufacturers in order to outfit the new British armory at Enfield, near London, which was to be operated on the "American System." Through the exhibits of Colt and the firm of Robbins and Lawrence (of Windsor, Vermont) at the Great Exhibition of 1851, in London, the idea of the "American System" had been brought home to English observers, with the results just indicated.

Thus we have examples that epitomize the mechanical "know-how" of Americans in 1800 and at mid-century, both in the production of large and small products. A number of distinguished studies have been made of canal, railroad, steamboat and manufacturing developments, but the concern has been with economic rather than technological development, and it is easy for a reader to assume that not much beyond Yankee ingenuity was necessary to build the locomotives, steamboats and manufacturing machines, so long as the money could be found to pay for them. Simple economic pressures and Yankee ingenuity are insufficient, however, to explain how a people could not only refine manual skills and improve upon machine tools that found their way to this country from the old world, but could strike out on a boldly original tack and, within a space of less than two generations, begin to export to England a manufacturing know-how and machine tools so different from those in England that the whole performance became known abroad as the "American System." Not only were manufacturing techniques and tools being exported, but in the 1840s American locomotives were being shipped to England, and American locomotive builders had gone to Russia and Austria to set up shops to build locomotives for the governments' railroads.

All of this occurred in spite of the fact that English machine-tool development started at least a generation ahead of American development. John Wilkinson's boring mill of 1775, which was a decisive factor in the success of Boulton and Watt's steam engine, was a much more sophisticated tool than Nicholas Roosevelt's boring mill of 1800.

When, in 1807, Robert Fulton wanted an engine to propel his pioneer steamboat, he imported a Boulton and Watt engine from England. It is significant, however, that an act of Parliament was required in order to export the engine, for an embargo on machinery of all kinds, intended to suppress foreign competition with English manufactured products which was not lifted until after 1840, was then in effect. This embargo, as well as the unavailability of British goods during the War of 1812, had a positive effect in encouraging the development of American tools and American machines, whose design may sometimes have departed from British precedent because the precedent was not at hand to be observed. The embargo was one factor certainly. But we ought to know something about the kind of people who were doing work in the United States, and about their sources of technical information and the ease with which it could be obtained. This has several ramifications, such as publications, technical societies, observations of travelers and so forth.

In trying a few years ago to learn something about what innovations actually occurred in the United States, I found this period to be a virtual wasteland, so far as actual source material on techniques and tools was concerned. The products of craftsmen and of machine tools could be traced without much difficulty, but the basic technical information was elusive and unsubstantial. There are, to my knowledge, no American machine tools of earlier than about 1830 in existence, with the notable exception of the Blanchard gun-stock machine of 1820, now preserved in the Springfield Armory Museum, and an incomplete device that looks like a milling machine, attributed to Eli Whitney but of quite uncertain origin. Nor have many drawings been preserved that can supply useful information. There are two reasons for this. First, many machines were built from full-size drawings on chalkboards or from no drawings, and second, even when drawings existed they were seldom considered suitable documents to be collected by libraries and historical societies.

At length, however, I ran across "An Interesting Letter from an Old Engineer" in the first volume of *Machinery*, of 1896. This letter, written by George Escol Sellers, led me back to a series of some 40 articles that Sellers had written for *American Machinist* between 1884 and 1893. These articles, "Early Engineering Reminiscences," began to fill the vacuum so far as information was concerned, and to bring the whole period alive and to provide plausible answers to many of the questions that I had asked. I will say that this was a most unlikely source, because it was written by a man who was past 75 before his first article appeared, and some of the events that he described so vividly had occurred before he was 10 years old. But I have tested the information, nearly line by line, and it has held up astonishingly well. In fact, I have thought enough of its value as source material to edit and annotate the series for publication. It is in the process of checking Sellers' statements that I have been able to recognize the significance of fragments whose relevance would not otherwise have been evident.

George Escol Sellers was born in Philadelphia in 1808, about a block and a half from the old State House, now known as Independence Hall. One of his grandfathers was Nathan Sellers, who was the first maker of wire moulds for handmade paper in the United States. Nathan's son, George Escol's father, followed in the mould-making trade, but branched out into the manufacture of fire engines and, eventually, paper-making machinery. The other grandfather was Charles Willson Peale, portrait painter to the great federal generation of Americans and founder of the museum that became, in its way, an educational institution for American mechanicians.

George Escol grew up in a mechanical household, and he was at home in the shops of a remarkable group of able craftsmen who worked in Philadelphia during the first half of the nineteenth century. He attended the Friends' School on Fifth Street until he was 15 or 16 years old. His classmates included Solomon and William Milnor Roberts, who became civil engineers, John Dahlgren, or naval ordnance fame, and John Trautwine, whose later civil engineering handbook became a classic in its field. Sellers attended a mechanical drawing class conducted by William Mason, a machinist and instrument maker, and he drew for John Haviland, the architect. He served no apprenticeship as such, but he became a competent machinist, profiting in every way by his associations with such fine craftsmen as Isaiah Lukens, Joseph Saxton, his uncle Franklin Peale, and an itinerant German aristocrat identified thus far only as Henri Mogeme.

The knowledge that a group of craftsmen, however competent, could impart to a pupil was limited. The existence and availability of printed information was important then as it is now. Oliver Evans (1755–1819), who was acutely aware of this need, figured in some of Sellers' earliest recollections.

Evans, a leading machine-builder in Philadelphia, was a friend of the Sellers family. A gifted innovator, Evans originated before 1800 an automatic flour mill which employed bucket conveyors, screw conveyors and automatic bolting equipment. Before 1810 he had built some of the first high-pressure steam engines in the United States and had invented a straight-line linkage that still bears his name throughout the world including Russia. He published the *Young-Mill-Wright and Miller's Guide* in 1795, more than ten years before the first similar work appeared in England. The *Guide* went through fifteen editions, the last appearing in 1860. Nevertheless, the title of his 1805 book, *Abortion of the Young Steam*

Engineer's Guide, reflected his frustration at the indifferent support he was able to command for a pioneering work on the steam engine which was issued before any similar work had appeared in England. Sellers told of Evans' concern over "the difficulties inventive mechanics labored under for want of published records of what had preceded them and for works of reference to help the beginner." *The North American Review*, in 1819, corroborated this deficiency:

> Books and instruments connected with the science and practice of engineering [are not] possessed by individuals in great numbers, and if any person seeks for them in the shops or book-stores in the United States, he will be disappointed. He must import them for his own use at great expense.

Although there were two or three mechanical periodicals in England at this time and some issues at least were in the Philadelphia Library Company collections, it was not until the 1820s that any considerable amount of technical information became available. In 1824, the Franklin Institute was formed in Philadelphia for the exchange and dissemination of technical information. Mechanics' Institutes were formed in New York, Boston, and Baltimore during this decade also, and in all four cities exhibitions of mechanical products, sponsored by the institutes, were held periodically. The Franklin Institute was easily the most important of the institutes, as measured by its influence upon the mechanical community. Its "Committee on Science and the Arts" acted as a clearing house for patent information and had the courage to judge the merits of individual inventions. Later committees on boiler explosions and their prevention published their findings for the benefit of all who would read. The Institute's *Journal* which appeared first in 1825, was equal in quality to Newton's *Journal* and the *Mechanics' Magazine*, both of London, which preceded it by a few years. Also, in 1823 an American edition of Abraham Rees' outstanding English *Cyclopaedia* had been completed.

Whatever his sources of information, the native-born American mechanician made significant contribution to the American tradition during this period. We should know more about these men than merely their names, although it is sometimes difficult even to learn names. Person accounts are especially useful in putting flesh on bare bones. The Sellers reminiscences offer a delightful as well as plausible catalogue of the mechanics that he knew.

Jacob Perkins (1765-1838), of Newburyport, Massachusetts, was a prolific inventor who in 1815 came to Philadelphia. His head fairly rattled with ideas; he kept his head covered by a stove-pipe hat in which he carried all manner of notes, sketches and memoranda. In 1819 Perkins emigrated to England, seeking a wider field for his talents and to introduce there a system of bank-note engraving that he considered proof against counterfeiting. He took with him Asa Spencer, another native of Massachusetts, who had originated an ingenious geared scroll-lathe to produce the intricate scroll designs required for bank-note engraving plates. Perkins' inventions covered the mechanical field: extremely high-pressure steam engines and boilers, a hot-water heating system, a vapor-compression refrigeration system, among many others. Although Perkins remained in England for the rest of his life, he left in the United States a legacy of ideas and enthusiasm for innovation, and his influence upon later American visitors to England was not inconsiderable.

Perkins was characterized by Sellers as a bustling, quick, enthusiastic and excitable man of boundless energy. He had the faculty of keeping the mechanical world "in a feverish state of excitement. . . . It was never what he had done but what he was doing." The fact that few of his schemes ever actually succeeded seemed to matter little when he looked about for money to support some new venture. "To sum it all up," wrote Sellers, "he certainly filled a useful place in advancing improvements in steam engines, for his schemes set many level headed men to thinking in the right direction."

Contrasted with Perkins was the builder of fire engines, Patrick Lyon (1769-1829), who is known today mainly through John Neagle's handsome painting of "Pat Lyon at the Forge." Lyon was a careful, expert craftsman in the iron and brass work of fire engines. Incidentally, two of his fire engines can be seen on display in Independence Hall. One is attributed to an earlier maker, but the mechanical details are so similar to Lyon's work that the attribution is almost certainly in error, for Pat Lyon was never a man to copy blindly. He was independent, confident in his own solid ability. When questioned about tariff protection in the federal census of 1820, he replied, even though his business had fallen off in the depression of 1819, "I manufacture cheaper and better than the articles I manufacture can be imported. I do not want any additional duty laid for my protection." He represented the intelligent, systematic mechanician who advanced the art by small but sound improvements.

William Mason and Rufus Tyler were two machinists who set the tone for excellence of workmanship. Fine turning lathes, including rose-engine lathes, and machinery for engraving calico-printing plates and bank-note plates were made in their shop. When, around 1822, a Maudslay lathe slide rest that had been smuggled out of England was brought to their shop, Mason and Tyler made changes and essential improvements, producing a sturdy and widely-accepted American version of the slide rest. As I have already mentioned, Mason taught youngsters how to use drawing instruments in mechanical drawing. Although Sellers did not tell whether Mason and Tyler had any apprentices in their shop, it is likely that they did.

Matthias Baldwin, world-famous builder of locomotives, was a Philadelphia mechanic who had served an apprenticeship with a jeweler and had spent many years as an engraver and fine machinist before a model that he was commissioned to build for the Peale Museum gave a new direction to his career. After the 1829 Rainhill trials of locomotives in England, Franklin Peale, at this time in charge of the museum, asked Baldwin to build for him a working model of the *Novelty*, the Ericsson and Braithwaite locomotive that had competed unsuccessfully with Stephenson's victorious *Rocket*. Why Peale chose the *Novelty* and not the *Rocket* as his model is not clear. However, it was for Baldwin but a step from the museum model to a full size locomotive for the Philadelphia and Germantown Railroad; and within a few years he was the leading locomotive-builder in America.

Franklin Peale was a good craftsman in his own right. He had served his apprenticeship in the machine shops of the Hodgson Brothers, located on the industrial stretch of the Brandywine Creek just north of Wilmington, Delaware. The Gilpin paper mills, the Young cotton factories, and the DuPont powder mills were among the Hodgsons' customers. Peale spent about two years in Europe for the U.S. Mint, studying the metallurgy and mechanics of minting processes throughout Europe. An extensive report of his is in the National Archives, but the hundreds of working drawings that he prepared from actual machinery in Europe apparently were not saved.

William Norris was a locomotive builder who was an organizer rather than a craftsman. After a short career as a dry-goods merchant, Norris set up a shop to build a locomotive designed by Stephen Long, an Army engineer, and earlier the discoverer of Long's Peak in Colorado. The trial of the first Long and Norris locomotive was described years later by Norris:

"Gentlemen, I can, on my honor, assure you that we ran four miles and a-half in seven hours and a quarter, and running all the time at that." Norris soon obtained the necessary mechanical help to eliminate his difficulties, and within a few years he built a locomotive that attracted attention in Europe as well as at home by climbing an inclined plane while pulling a loaded passenger car. Until this feat was accomplished, it was assumed that locomotive engines would pull a load only on nearly a dead level, and that considerable changes in elevation would have to be overcome by inclined planes, employing stationary engines to pull cars up. The idea had been imported uncritically from England where it had been stated by George Stephenson, one of the pioneer builders.

Norris locomotives were soon being exported to England; and it was Norris who in the 1840s went to Vienna to organize locomotive and car shops for the Austrian state railroad.

In the United States, a considerable number of English locomotives were imported for the early railroads. Given the demand for locomotives, however, specialized machine shops came into existence in Philadelphia, New York, New Jersey, and Massachusetts. George Escol Sellers and his older brother Charles built two locomotives for the Philadelphia and Columbia Railroad. Seth Boyden, of Newark, New Jersey, was a versatile inventor who turned to locomotive building. John Brandt, a Lancaster, Pennsylvania blacksmith who was concerned with the Sellerses for a time in building a textile-card making machine, went on to become master mechanic of the Erie Railroad and eventually to set up a successful locomotive works in Lancaster.

The know-how of the mechanicians just described was supported not only by the scanty technical literature available and specimens of English machines but also by immigrants who brought with them needed skills. Sellers' account of an incident that occurred in the 1820s, when his father was running an extensive fire-engine manufactory is revealing. A German workman came to the shops one day looking for work. After George Escol's father had told him that he had nothing for him, the German stood about for some time watching the work being done in the shop. At length he said, to no one in particular, "I see no coppersmith's bench or tools." Sellers said, "We have that work done out." The visitor countered, "Better work, better fits if done here."

The result was that the German, who gave his name as Henri Mogeme, built himself furnaces and tools required for a small brass foundry and fitting shop, and in the course of his work he found time to instruct George Escol in the art of brass-founding and brazing. Sellers said later, "I never went into his shop that I did not learn something. Mogeme was a capital teacher: when explaining anything to me it would be as if I was entirely ignorant. He would say, 'No understand without beginning right.'"

Mogeme was an itinerant craftsman, but with this difference. He apparently was of noble birth, had attended a German technical school and had worked in various shops on the continent and in England before coming to the United States. He returned home after a few years in America; but he left the country richer so far as mechanical skills were concerned. I have not heard of other such itinerants in the United States; but as I mentioned earlier the whole technical background of this period is very imperfectly known.

Still another source of know-how which is not generally realized is the American mechanician who went abroad in search of know-how on the spot. In 1832, when he was not quite 24, George Escol Sellers went to England to learn what he could about the building of paper-making machines and their use. The first paper machine in the United States was the one that had been installed in the Gilpin mill, on the Brandywine, in 1815, after both of the Gilpin brothers had visited England and after an English paper-maker, Lawrence Greatrake, had been induced to emigrate to America. The Gilpin machine was a copy of John Dickinson's English cylinder machine, patented in 1809. The Sellers shops had been for several years building a much-simplified version of the Dickinson cylinder, which enabled small paper mills to increase production with a trifling capital expenditure. In the late 1820s, however, an English Fourdrinier machine was imported and installed in a mill in upstate New York. The idea of the Fourdrinier machine had originated in France, but the machine development had been financed by the Fourdrinier brothers, London stationers, and the actual development work had been largely in the hands of Bryan Donkin, also of London. To see Donkin's shop was one of George Escol's primary objectives in going to England.

Sellers was entertained and treated with unusual openness by some of the outstanding mechanicians of the day, and he was thus able to visit several prominent shops and mills and to compare English with American practice. Although he was but a young man, the relative ease with which he gained the confidence of and obtained information from such cautious men as John Dickinson and Bryan Donkin suggests that Sellers was an intelligent—even expert—listener to detailed technical descriptions, and that the information that he had brought with him from America was of great interest. Since George Escol's sister was engaged to marry a man who had come from Birmingham, England, the bridegroom's brother took Sellers in hand when he arrived in England and opened several doors for him; and his good friend Joseph Saxton, a Philadelphia mechanician who had been in England for some time, was his constant companion in London, introducing him to his circle of acquaintances.

Sellers was greatly impressed by Donkin's accomplishments. He noted that while he was in his company, one day, Donkin received word of a Fourdrinier machine that had broken down. Taking spare parts from the shelf, Donkin dispatched a workman to get the machine back in order, predicting that it would be making paper again before midnight. The idea of an interchangeable spare part for a musket was gaining general acceptance by this time in the United States, but a spare part for a machine as extensive and complex as the Fourdrinier paper machine was unheard of.

On the other hand, the weight of traditional practice sometimes blinded the great English craftsmen to the value of American improvements. Donkin was interested in what Sellers told him about mechanical developments in America, but he could not understand how machine works could be successfully run without a strict division of labor, as in England, where each workman mastered only a single operation, such as turning, filing or wielding a cold chisel. "I told him," wrote Sellers, "that he must bear in mind that America's start in mechanical art was at the point England had reached and without her prejudices." One example that Sellers cited was England's rejection of the American cut nail, and the prejudice of English carpenters and patternmakers for the handmade tapered wrought nail, of "the best possible form that could be devised to split the wood it was driven into, without first boring a hole to receive it." Another example was the pointed wood-screw, which originated in the United States. The Birmingham hardware makers would not consider taking orders for such screws, because the old square-end screw was good enough for them.

Sellers had also an opportunity to visit and inspect at leisure the celebrated shops of Maudslay, Son and Field, machinists and engine builders. Henry Maudslay, the pioneer machine-tool builder, had died in 1829; the works were being carried on by his partners. Sellers was taken first by his host to see the boring machine on which the largest steamboat engine in England was being machined.

"Here," wrote Sellers, "I must confess a feeling of great disappointment, for the cylinder struck me as a mere pigmy compared with the cylinders of the North River and Long Island Sound boats of that period. . . . There were at that time boats on the American rivers with condensing engines, whose cylinders would cover two of the one on the boring machine. . . . The workmanship on them appeared to be of the highest possible character. This astonished me, after having seen the lathes and other machine tools, none of which lacked in care or accuracy in their construction, but totally inadequate for the character of the work they had to do, as to weight, strength, and firmness." Moreover, none of the lathes he saw was as large as the one his father, his brother and he were then (in 1832) building in their Philadelphia shops.

The British, however, were by no means unaware of American advances, and an opportunity to look at American development from an English viewpoint is afforded by British Parliament inquiries concerning machinery —one in 1841 regarding the operation of the machinery embargo and another in 1854 regarding the making of muskets.

An American who was in London in 1841 answered a question about the use of English tools in the United States. "There are not a great many English tools imported into the country," he said, "but the English have furnished us with some important patterns to build by. The planing-machine we first obtained from England, and I believe the first imported into the United States was smuggled from this country, from the manufactory of Messrs. Sharp, Roberts & Company, of Manchester."

An English witness noted, on the other hand, that the "entirely new inventions; not improvements in machines, which are still mostly made here; come now from abroad, especially America."

Another Englishman, who had been in the United States, recalled that the machinery there was "tolerably well finished, such as we should call second-rate machinery in our own country."

In answer to the question: "Are the Americans as skilled as the English?" this same witness replied, "It is not possible that they can be, without more experience than they have had at present."

"Are the Americans aware of their inferiority?"

"They will not allow it," responded the witness.

David Stevenson, Scottish engineer and uncle of Robert Louis Stevenson, had reported elsewhere that he had seen in the Baldwin locomotive works in Philadelphia "no less than twelve locomotive carriages in different states of progress, and all of substantial and good workmanship. Those parts of the engine, such as the cylinder, piston, valves, journals, and slides, in which good fitting and fine workmanship are indispensable to the efficient action of the machine, were very highly finished, but the external parts, such as the connecting rods, cranks, framing, and wheels were left in a much coarser state than in engines of British manufacture."

In 1854, when the "American system" was being discussed by a Parliamentary committee, Samuel Colt was called in to testify. He had much to say about the "American system," but he managed to express the American position in a sentence: "There is nothing [he said] that cannot be produced by machinery."

While Joseph Whitworth, a leading English engineer and mechanician, could see no particular advantage in the "American system" of manufacture, another prominent engineer, James Nasmyth, was generous in his praise of the system as it had been carried out in the Colt pistol factory. "The first impression," he said, "was to humble me very considerably. . . . In those American tools there is a common-sense way of going to the point at once, that I was quite struck with; there is a great simplicity, almost a quakerlike rigidity of form given to the machinery; no ornamentation, no rubbing away of corners, or polishing; but the precise, accurate, and correct results." Whitworth, who had visited the United States in 1853, said elsewhere that the thing that had impressed him was the ready acceptance of labor-saving tools by the workmen as well as the proprietors.

☆   ☆   ☆

A number of the elements of answers to questions implicit in the title of this article are contained in the sources and lines of inquiry that I have suggested here. The catalogue is by no means complete, but I have ventured to give a fragmentary account at this time in the hope that others will perhaps be provoked to pursue the subject further.

I close with a few tentative conclusions, about which argument or discussion are invited.

First, the information that came from Europe was essential, and it was used freely and without prejudice. Next, the stream of travelers going to

Europe to obtain mechanical and engineering information was important, but its magnitude is not known with any precision. I suspect that it was perhaps five times as great as the best-informed scholars today would estimate it to be.

Third, and this I have not illustrated because it is adequately covered elsewhere, the geography, unlimited natural resources, economy and political climate of the new country all had a powerful influence upon mechanical developments. Governments at all levels were permissive but not, as we have been generally taught to believe, passive. Patent laws, corporation charters, government subscriptions to stocks, and direct subsidies have served to promote the common good (at least in the production of material wealth) by encouraging individual ingenuity and enterprise.

Fourth, the intelligence, ability and self-reliance of the mechanics mentioned here—and there were many hundreds of others like them—was certainly an important factor in the development of the tradition of "know-how." These men, while often short on principles, possessed a highly developed intuitive sense of fitness, as well as an integrity that insured an honest product of good value. There were a few catalysts, such as the Perkinses and the Ericssons, who added suggestion if not direction to the mainstream of engineering advance.

A final ingredient, found in the United States but not in Europe, was the freedom with which knowledge was shared and exchanged. Closed shops and mills seem to have been few and far between in the United States. To show conclusively that this "open door" policy was completely general will require more information than I now have; but when the factors that underlie the rise of American mechanical "know-how" are ranked in order of importance, I think that this one will be close to the top of the list.

The Heroic Theory of Invention

LOUIS C. HUNTER

Few inventions have been as important as the steamboat, which had a revolutionary effect on western development. It was also the means by which the steam engine was introduced to America. The arrival of the steamboat signaled the beginning of a new age; but, like many technological innovations, its origins have been obscured by the heroic theory of invention, which explains important developments in terms of a single genius. Louis C. Hunter shows that the western steamboat was not the creation of any single man; it was an evolutionary development to which many men contributed. Hunter, an economic historian, is the author of *Steamboats on the Western Rivers* (Cambridge, Mass., 1949), a classic in the history of technology. Now retired, Dr. Hunter was professor of history at American University; he is currently preparing a history of industrial power in the United States.

I

The steamboat was the first great American contribution to modern technology. First developed to the point of practical commercial success in this country, the steamboat was quickly incorporated into the economic structure of the nation and within a few years became the principal vehicle of transportation on the main channels of inland commerce. Eventually the influence of the American steamboat was felt all over the globe, and there were few great river systems that in time did not feel the beat of the paddlewheels by which it was propelled.

Louis C. Hunter, "The Invention of the Western Steamboat," *The Journal of Economic History*, 3 (November 1943), 201–220. Copyright © 1943 by the Economic History Association. Reprinted by permission.

That the individual credited with "inventing" a device of such influence and fame should be raised to a high pedestal and ranked among the elite of the nation's heroes is natural enough. In the popular history of American technology Robert Fulton occupies a place comparable to that of James Watt in England. The device that he fathered no longer plays as important a role in the internal commerce of the nation as it once did, but the steamboat tradition nevertheless has a firm hold on the American imagination. In the narrower world of scholarship, however, Fulton's position has in recent years been badly shaken. The assumption that he was solely or principally responsible for inventing the steamboat has been attacked from many directions, chiefly by biographers intent on asserting the claims of rival contemporary experimenters. So effectively have the claims of such men as Stevens, Fitch, Rumsey, and Oliver Evans (to mention only the leading contenders) been put forward that the position of Fulton has been reduced, at best, to that of *primus inter pares*.[1]

While Fulton's role as the inventor of the steamboat has been undergoing attack from biographers of his eastern rivals, another group of historians has been undercutting with no less determination his position as the father of steamboating on the western rivers. It will be recalled that within five years of his initial success on the Hudson with the *Clermont*, Fulton, in association with Livingston, Roosevelt, and others, introduced steam navigation on the Mississippi-Ohio river system. Beginning with the *New Orleans* in 1811, this eastern group of promoters built and put into operation several steamboats of substantial tonnage. An ambitious plan was projected of providing steamboat service along the whole length of the rivers between Pittsburgh and New Orleans. Grants of exclusive privileges were therefore sought from the States along the way. While this plan was realized only in part, enough was accomplished to found the tradition that Fulton and his associates were the pioneers who led the way in the development of a system of river transportation which in the nineteenth century was without parallel in other parts of the world. In the western, as in the eastern sphere, Fulton has therefore long been awarded the fame which, in a brand of history pervaded by the

---

[1] *See* especially A. D. Turnbull, *John Stevens, an American Record* (London: The Century Company, 1928), Thomas Boyd, *Poor John Fitch* (New York: G. P. Putnam's Sons, 1935), Greville and Dorothy Bathe, *Oliver Evans* (Philadelphia: The Historical Society of Pennsylvania, 1935).

heroic conception of invention, is customarily bestowed upon an individual first in the field.

With this interpretation, which pictures steamboat transportation on the western rivers simply as an extension of eastern ingenuity and enterprise and which casts an eastern man in the role of hero, a number of historians, westerners to be sure, have taken sharp issue. Indeed, they have attacked the Fultonian view so vigorously that it has been generally superseded in works dealing with this phase of western development. It is not denied that Fulton and his associates introduced the first steamboats on the western rivers; in fact, a certain measure of credit is usually conceded to the eastern group for their labors in the western field. But the importance of the contributions of Fulton is minimized chiefly on the ground that his western steamboats were technically a failure, and that Fulton contributed virtually nothing to the solution of the basic problems of steamboat navigation on the western rivers. Whatever stimulus to steam navigation the Fulton group may have given was offset by the deterrent influence of their attempt to enforce monopoly claims, based on the Orleans grant of exclusive franchise. It could be argued, therefore, that the net result of the activities of Fulton and his associates in the West was to delay rather than to hasten the development of steamboat transporation in his region. Yet the critics who challenge Fulton's claims to honor, in keeping with the traditional conception of technological change, pull Fulton off his pedestal only to replace him with another hero, a Westerner, Henry Miller Shreve.

II

The case for Henry Shreve as the father of western steamboating rests on several grounds: his successful challenge of the Fulton-Livingston monopoly on the lower Mississippi; his activities in the field of river improvement, particularly his invention of the snagboat; and his contributions to the structural and mechanical development of the western river steamboat. The last item is the principal basis for Shreve's fame and to it I shall direct chief attention. Shreve's career on the rivers dates from the keelboat age when, as boat owner and captain, he made numerous voyages up and down the Ohio and Mississippi rivers, particularly in the Pittsburgh-New Orleans trade. In 1814 he became associated with the group of men at Brownsville, on the lower Monongahela in western Pennsylvania, who offered the first challenge to the Fulton-Livingston plan for monopolizing steamboat transportation on the western rivers. In 1813 this group built the second steamboat in the West, the *Comet* (25 tons), and the following year the *Despatch* (25 tons) and the *Enterprise* (75 tons), all equipped with the type of steam engine invented by Daniel French, an eastern mechanic who had settled in Brownsville. The *Comet* was a failure, but the *Despatch* did moderately well, while the *Enterprise* was quite successful, making several notable voyages under the command of Shreve, her part owner. In 1816 Shreve became part owner of a new steamboat built according to his specifications and equipped with machinery of his own design. This was the *Washington* (403 tons), celebrated in western tradition as the first great steamboat on the western rivers. The third of the early steamboats with which Shreve's name and fame have been intimately associated was the *George Washington* (356 tons), built in 1824–1825 and embodying important innovations.

In the *Washington* and the *George Washington*, so runs the now widely accepted account, Shreve established the prototype of the western river steamboat. In their design and equipment these vessels broke radically with the type of vessel introduced by Fulton in the West and thereby assured the success of the steamboat in this region. As one writer on the history of western steamboats recently declared, Shreve "produced the ideal vessel to navigate Old Miss and her tributaries—a unique vessel without which the West would never have been "the West." That ideal vessel was the steamboat and Henry Shreve created it."[2] The most prolific historian of western river transportation wrote in similar vein: "The task of constructing a great inland river marine to play the dual role of serving the cotton empire and of extending American migration and commerce into the trans-Mississippi region was solved when he [Shreve] built the *Washington* at Wheeling in 1816. . . . The remarkable success of his design is attested by the fact that in two decades the boats built on his model outweighed in tonnage all the ships of the Atlanic seaboard and Great Lakes combined."[3] Still other writings on steamboat history refer to Shreve as "the man who was chiefly responsible for the steamboat as it developed

---

[2]G. L. Eskew, reviewing Florence L. Dorsey, *Master of the Mississippi* (Boston: Houghton Mifflin Company, 1941) in New York *Herald Tribune Books*, November 9, 1941.

[3]A. B. Hulbert, *The Paths of Inland Commerce* (New Haven: Yale University Press, 1920), 174–175.

on the western rivers" and as "the real creator of the American steamboat."[4] Particular weight is given to Shreve's achievement by alleging the failure of the Fulton boats to overcome the difficulties of western river navigation. It was the original intention of the eastern group to provide steamboat service all the way from Pittsburgh to New Orleans. But, so runs this story, their vessels proved to have too shiplike a design and too deep a draft to operate on the upper parts of the river system; moreover they lacked the power to overcome the swift currents encountered on the upstream passage from New Orleans to the Ohio River ports. Consequently the Fulton boats spent most of their time running in the New Orleans-Natchez trade on the lower Mississippi. Since the great and primary need of the West was for a cheap and rapid means of upstream transportation, the conclusion follows that, in the larger sense, the Fulton boats were a failure. But whereas Fulton failed, Shreve succeeded; and western commerce and settlement advanced as in seven-league boots. As captain of the *Enterprise*, Shreve commanded the first steamboat voyage up the river from New Orleans to Louisville, made in 1815. While Shreve shared in the glory of this achievement, the allegedly exceptional conditions under which this voyage was performed—a flood stage so high as to enable the pilot to "cut across lots" and avoid the full force of the current—in a vessel not of Shreve's design left the door open for a triumph that could be wholly his. The public, we are told, continued to be skeptical of the steamboat's ability to stem the swift currents of the Ohio and Mississippi under ordinary conditions.

These doubts of the public were put to rest when two years later Shreve made the long, hard trip from New Orleans to Louisville in the *Washington*, the vessel which he himself had designed, without the aid of favoring conditions and in the fast time of forty-five days, including the time in port, for the round trip and twenty-five days for the upstream voyage. "This was the trip," declared M'Murtrie, a Louisville citizen writing in 1819, "that convinced the despairing public that steamboat navigation would succeed on the western waters."[5] And this was the statement that, quoted and paraphrased repeatedly in later years, with little or no corroboration, convinced western writers on steamboat history and pro-

[4] M. L. Hartsough, *From Canoe to Steel Barge on the Upper Mississippi* (Minneapolis: University of Minnesota Press, 1934), 44; G. L. Eskew, *The Pageant of the Packets* (New York: Henry Holt and Company, 1929).

[5] H. M'Murtrie, *Sketches of Louisville* (Louisville, 1819), 202.

vided one of the main supports of the Shreve legend. "After this memorable voyage of the *Washington*," wrote Lloyd a generation later, "all doubts and prejudices in reference to steam navigation were removed. Shipyards began to be established in every convenient locality, and the business of steamboat building was vigorously prosecuted."[6] According to Dunbar, the population of the Mississippi Valley was as excited over the achievement of the boat designed and captained by Shreve as over Jackson's victory at New Orleans and as the news spread through the Ohio Valley the construction of numerous steamboats was begun.[7] Other scholars accepted this view of the critical importance of the famous voyage of 1817.[8]

III

Useful as this interpretation is in building the life of a hero, it does not stand up well under critical examination. At least three steamboats had made the trip up the rivers from New Orleans prior to the *Washington*. Two of them were steamboats constructed by the Brownsville group, equipped with engines designed and built by Daniel French. In 1815 the *Enterprise*, a much smaller boat than the *Washington*, made the first trip up the river in the fast time of twenty-five days.[9] Even if we accept the version of this voyage that minimizes its importance—and there are good reasons for questioning it[10]—the solid fact remains that this vessel

[6]James T. Lloyd, *Lloyd's Steamboat Directory and Disasters on the Western Waters* (Cincinnati, 1856), 45.

[7]Seymour Dunbar, *A History of Travel in America* (New York: Harlem Publishing Company, 1937), 395.

[8]*See*, for example, F. F. Gephart, *Transportation and Industrial Development in the Middle West* (New York: Columbia University Press, 1909), 72–73; B. H. Meyer et al., *History of Transportation in the United States before 1860* (Washington: Carnegie Institution, 1917), 104; A. B. Hulbert, *The Ohio River* (New York: G. P. Putnam's Sons, 1906), 334–335; Leland D. Baldwin, *The Keelboat Age on Western Waters* (Pittsburgh: University of Pittsburgh Press, 1941), 192.

[9]*Cincinnati Western Spy*, April 19, 1816.

[10]I have found no contemporary evidence in support of the view expressed above that the *Enterprise* was aided by flood conditions. Against this view must be placed the well-known facts that navigation at flood seasons was more difficult than at an ordinary stage, due to the greater swiftness of the current and the presence of great quantities of driftwood. While the distance covered could be shortened somewhat through the use of cutoffs and otherwise impassable island chutes, the force of the

ascended not only to Louisville but 600 miles further upstream to Pittsburgh and then some fifty miles up the Monongahela to the home port of Brownsville. In the contemporary press the feat of the *Enterprise* received attention and favorable comment comparable to that received by the *Washington* two years later.[11] In 1816, the diminutive *Despatch* also made the difficult upstream voyage, completing the trip from New Orleans to Louisville in thirty-four days.[12] Prevented from taking cargo by the representatives of the Fulton group at New Orleans, the *Despatch* did not receive on this trip a full test of her ability to stem the current of the Mississippi.

To the achievements of the French-engined steamboats there were soon added those of two of the Fulton boats, the *Vesuvius* (340 tons) and the *Aetna* (360 tons). The *Vesuvius* started up the river for Louisville in the spring of 1814 but failed to complete the trip, thereby adding weight to the charge that the Fulton steamboats were a failure. Since the contemporaneous account of this voyage does not make clear the cause of failure, whether personal, mechanical, or architectural, judgment must be reserved.[13] Later, however, the *Vesuvius* was to redeem herself. But there can be no blinking the achievement of the *Aetna*, since, prior to the "epochmaking" voyage of the *Washington*, the *Aetna* had completed three trips up the river from New Orleans to Louisville. The first was made in the autumn of 1815 when she had the misfortune, not uncommon in the early years, to break a shaft, not far from the mouth of the Ohio. After spending fifteen days in an unsuccessful attempt to make repairs, her officers succeeded in completing the voyage on one paddlewheel, no mean feat in itself, in a port-to-port time of sixty days.[14] In the spring of 1816 the

---

current could not be avoided to a much greater extent at flood seasons than at other times, because of the frequent necessity of crossing from one side of the river to the other.

[11] *Cincinnati Western Spy*, October 6 and November 24, 1815; *Niles' Weekly Register*, July 1, 1815; *Lexington Reporter*, September 6, 1815; *The American Telegraph* (Brownsville, Pa.), September 20, 1815.

[12] *Cincinnati Western Spy*, April 19, 1816.

[13] According to this account the *Vesuvius* grounded on a sandbar 700 miles up the river from New Orleans about June 1 and remained there until December 3 when the river rose and floated her off. She then returned to New Orleans where she ran aground a second time on the Batture and lay there until released by a rise in the river about March 1. *Liberty Hall and Cincinnati Gazette*, August 3, 1819. The precise causes for these mishaps are subjects for conjecture.

[14] Ibid., January 15, 1816; *Niles' Weekly Register*, July 1, 1815.

Aetna made a second trip up the rivers to Louisville with a port-to-port time of thirty-five days but with an actual running time of but thirteen days.[15] A year later she arrived at Louisville from New Orleans for the third time, a week in advance of the *Washington* on her much acclaimed trip. If the *Aetna* on this occasion made much slower time than the Shreve-designed boat, her cargo was half again as large as the *Washington's* with a freight bill amounting to more than $25,000.[16] "Whilst the Steam Boats are in charge of such persevering and skilful officers as the Captains of the *Aetna* and the *Buffalo*,"[17] ran a Cincinnati editorial, "we need not fear of success in ascending the western rivers to any navigable point."[18] In praising his vessel, Captain Robeson de Hart of the *Aetna* remarked to the reporter that the round trip to New Orleans from Louisville could now be made in thirty-five to forty days.[19] Altogether, the *Aetna* made six voyages from New Orleans to Louisville under Captain de Hart and was continued in the trade under his successor.[20] During 1817 at least three other steamboats arrived at Louisville from New Orleans: the *Franklin* in thirty-six days; the *Buffalo* in an unreported time; and the *Vesuvius*, rebuilt after damage by fire, two trips, one of them in thirty-two days. Carrying a full cargo, the *Vesuvius* demonstrated again the ability of a Fulton-designed boat to master the currents of the Mississippi and Ohio rivers.[21]

Clearly the disparagement of the achievements of the Fulton group in the West, colored as it has been by resentment of their attempt to enforce their exclusive franchise on the lower Mississippi, has gone too far. From the fact that the operations of the Fulton boats were confined largely to

[15] Ibid., April 29, 1816.

[16] *Cincinnati Western Spy*, April 18, 25; May 2, 1817; the weight of the *Aetna's* cargo has been estimated from the freight bill.

[17] The *Buffalo* had just arrived on her maiden trip from Pittsburgh.

[18] *Cincinnati Western Spy*, May 2, 1817.

[19] The now traditional account of the record voyage of the *Washington* describes the public dinner given Shreve at Louisville in celebration of the event but fails to mention that Captain de Hart was also invited. "These two enterprising men," declared an editorial on the occasion, "have gained much of public esteem by their successful and enterprising exertions to demonstrate the practicability of navigating the Ohio and Mississippi, the high seas of the western country, with steam vessels." Ibid., May 9, 1817.

[20] *Liberty Hall and Cincinnati Gazette*, August 3, 1819.

[21] *Liberty Hall and Cincinnati Gazette*, June 16, 1817; *Cincinnati Western Spy*, June 27 and July 25, 1817; *The Reporter* (Lexington, Ky.), August 27 and December 10, 1817.

the lower Mississippi, those unsympathetic to the eastern group have too readily assumed that these vessels were unable to overcome the difficulties of the long upstream trip to Louisville. Actually they have overlooked other considerations which doubtless underlay this policy. As a glance at the population-distribution maps of 1810 and 1820 shows plainly, the portion of the lower river included in the Natchez–New Orleans trade was precisely that serving the region of greatest population density in the entire West outside of the Ohio Valley.[22] That the Fulton interests should have given first attention to exploiting the rich traffic of this easily navigated 250-mile stretch of deep, even-flowing water was natural enough. From Natchez to Louisville, on the other hand, for a distance of 1,000 miles by river there was little but frontier wilderness with a thinly scattered population of backwoodsmen and no towns or cities of importance to supply traffic and support for the new mode of transportation, to say nothing of assistance in case of accidents. Why undertake this long and difficult voyage through waters made dangerous by swift currents, narrow twisting channels, snags, and other obstructions when there was ample business to be done at a good profit on the deep and snag-free waters of the lower Mississippi?

More than financial and technical considerations were involved. If the terms of the Orleans (Louisiana) franchise did not expressly require that at least one steamboat be kept in operation chiefly on the waters of this territory, a certain sense of obligation on the part of the grantees may well have been an influence in this direction.[23] At any rate, one boat was kept on the New Orleans–Natchez run from the beginning. Whatever plans Fulton and his associates may have had originally for the extension of service up the rivers to Louisville and beyond were interrupted by a series of misfortunes which left them with but one steamboat in operation in the West during most of the period prior to the completion of the *Aetna* in 1815. By the time their second boat, the *Vesuvius*, was ready for operation in 1814, the *New Orleans* was sunk by accident and two years passed before her machinery was installed in the second *New Orleans*.[24] This vessel was hardly more than placed in service when a second calamity oc-

[22] J. K. Wright (ed.), C. O. Paullin, *Atlas of the Historical Geography of the United States* (Washington and New York: Carnegie Institution and the American Geographic Society, 1932), Plate 76.

[23] *Acts Passed at the Second Session of the Third Legislature of the Territory of Orleans* (New Orleans, 1811), ch. xxvi.

[24] *Cincinnati Western Spy*, February 2, 1816.

curred, the burning and sinking of the *Vesuvius*,[25] and it was almost a year before she was salvaged and rebuilt. A fourth steamboat in the original series, the *Buffalo*, was never completed by the Fulton group; financial difficulties encountered during construction resulted in her sale under the sheriff's hammer.[26] With the second *New Orleans* taking care of the business in the New Orleans–Natchez trade, the *Aetna* and the rebuilt *Vesuvius* were made available for operation in the New Orleans–Louisville trade. Thus there is no need to resort to the theory of technical deficiencies to explain the confinement of the operations of the Fulton boats chiefly to the New Orleans–Natchez trade and the lower Mississippi.

The marked upswing in steamboat tonnage on the western rivers in 1818 and 1819 is usually attributed to the record voyage of the *Washington* in 1817, which allegedly banished all doubts of the practicability of steamboats. The experience of the year 1817 did, to be sure, promote confidence in the new mode of transportation, but this resulted from the achievement not of one but of a number of steamboats. Moreover, the public was interested in something more than evidence of technical success. The figures that carried greatest weight with potential investors, one may feel sure, were not so much those of a few days saved on a long trip as the big profits reported for a number of the pioneer steamboats. According to a detailed account of the *New Orleans's* first year, given wide publicity in the 1814 and later editions of the popular river guide, *The Navigator*, the owners cleared $20,000 over and above expenses, repairs, and interest on investment on a property valued at $40,000, a revenue, commented the editor, "superior to any other establishment in the United States."[27] It was reported that the *Enterprise* in the 1815 season would clear 40 per cent on her first cost; and 40 per cent, as a Cincinnati writer remarked, "speaks plain to every understanding."[28] A single trip of the second *New Orleans* from New Orleans to Natchez in 1817 netted a profit of $4,000.[29] In the same year the little *Franklin*, on her trip from

[25]Ibid., August 2, 1816.

[26]*Liberty Hall and Cincinnati Gazette*, August 3, 1819.

[27]Zadok Cramer, *The Navigator* (Pittsburgh, 1814, ed.), 31–32; 1817 and 1818 eds., 30–31.

[28]Communication on "Steam Boats" in the *Cincinnati Gazette*, reprinted in *The American Telegraph* (Brownsville, Pa.), September 20, 1815.

[29]Felix Flugel (ed.), "Pages from a Journal of a Voyage Down the Mississippi to New Orleans in 1817," *Louisiana Historical Quarterly*, VII, 436–437.

New Orleans to Louisville, cleared some $6,500.[30] The *Washington* paid her entire cost and divided $1,700 among her owners from the proceeds of two round trips between Louisville and New Orleans in the first half of 1817.[31] The handsome freight bill of the *Aetna*, on the 1817 trip referred to above, was greatly exceeded a year later by the $47,000 freight revenue of the *Vesuvius*, on a single trip in the same trade, half of which was said to be clear profit.[32] In describing a trip down the rivers in 1818, Estwick Evans declared that a voyage of a few weeks brought a return of 100 percent on the capital employed.[33] With reports such as these in circulation, the investors' rush into the new branch of enterprise is readily understood.

IV

The achievements of the Fulton boats have not received adequate recognition. Similarly, the contrast in design between them and Shreve's steamboats has been exaggerated. Writers who glorify Shreve assert that his *Washington* and *George Washington* were the prototype of the standard western river steamboat. The steamboats introduced by Fulton were essentially seagoing boats with deep and sharply modeled hulls and pronounced keels. With their heavy draft they were quite unsuited for use on the shallow waters of the western rivers. Shreve, we are told, quickly discovered the reason for the failure of these boats, radically altered their design and construction, and thereby made steamboats a practical success on these rivers. "Flagrantly ignoring the conventional wisdom of his day and craft, [he] built the *Washington* to sail *on* the water rather than *in* it, doing away altogether with a hold and supplying an upper deck in its place. To few inventors, indeed, does America owe a greater debt of thanks than to this Ohio river shipbuilder. A dozen men were on the way to produce a *Clermont* had Fulton failed; but Shreve had no rival in his plan to build a flat-bottomed steamboat."[34] Several writers have taken up and

[30]*Liberty Hall and Cincinnati Gazette*, June 16, 1817.
[31]*United States Magazine and Democratic Review*, XXII (1848), 169.
[32]W. Faux, *Memorable Days in America* (1823), 203.
[33]Estwick Evans, *A Pedestrious Tour . . . through the Western States and Territories, during the winter and spring of 1818* (1819), 154.
[34]Hulbert, *Paths of Inland Commerce*, 175; see also Hulbert, *The Ohio River*, 336.

repeated this theme: "The *Washington* . . . differed from its predecessors in that it had a flat, shallow hull. . . ."[35] "First of all she floated on the water and not in it."[36] "Her hull, having . . . a very shallow draft, sat on the water instead of sinking in it."[37]

What are the facts in the case? The fully developed western river steamboat had, it is true, a shallow hull and a flat bottom. Believing Shreve to have been the creator of this vessel, steamboat historians have concluded that these important features must have originated with him. There is, however, no contemporaneous evidence that even suggests that the hull of the *Washington* had a flat bottom or departed in other respects from the conventional model. On the contrary, descriptions by several contemporaries referred to her as "frigate built," "built like a ship," and "resembling a dismasted frigate."[38] There is no contemporaneous evidence yet discovered that gives the draft of the *Washington*, loaded or light, and none to support the assertion that she sat "on the water rather than in it."[39] All we have for the *Washington*, as for many of the pioneer western steamboats, are the hull dimensions as recorded in the customs-house enrollment documents, and these show a hull depth hardly compatible with shallow draft. We have but to compare the hull dimensions of the most famous of the Shreve boats with those of other vessels of their day to see how conservative a model was adopted by their designer. In so far as hull dimensions are indicative of design, neither of Shreve's famous boats diverged greatly from the pattern of their day. The much vaunted shallow hull of the *Washington* proves to have had what was probably the greatest depth of any vessel throughout the steamboat era. Such post-Civil War leviathans as the *J. M. White* (1,399 tons), *Thompson Dean* (1,368 tons), and *Grand*

[35] Article on Shreve in the *Dictionary of American Biography*.

[36] C. H. Ambler, *A History of Transportation in the Ohio Valley* (Glendale, California: The Arthur H. Clark Company, 1932), 127.

[37] M. L. Hartsough, *From Canoe to Steel Barge*, 45–46; *see also Baldwin*, 192.

[38] *Liberty Hall and Cincinnati Gazette*, September 23, 1816; "Reminiscences of J. Scott Elder," Louisville *Courier-Journal*, January 30, 1876; *Democratic Review*, XXII, 168.

[39] The foundation for this as for many other features of the Shreve legend was laid in the *Democratic Review* articles of 1848 upon which all histories of western steamboating lean heavily, although few authors appear to have consulted them directly. These articles, however, refer to the *George Washington* as the vessel in which Shreve broke with the traditional ship model. *Democratic Review*, XXII, 241–242.

HULL DIMENSIONS OF EARLY WESTERN STEAMBOATS[40]
(*in feet*)

Name	Year	Tonnage	Length	Breadth	Depth
*Washington**	1816	403	†13–.8	†2–.75	12.55
New Orleans	1811	371	148.5	32.5	12
United States	1819	645	. .	. .	11.25
Average, 400–500 ton class	1818	424	149	31.6	9.85
*George Washington**	1824–5	356	152.33	30.6	8.25
Average, 300–350 ton class	1827	318	143.8	26	8.89

*Shreve-designed boats.
†Figures not entirely legible due to mutilation of document.

Republic (1,794 tons) had hull depths of only ten and one-half feet.[41] The hull of the *George Washington*, so far as depth was concerned, represented only a modest improvement over other boats of her class. The case against the revolutionary influence of the Shreve boats is clinched by a study of the trend of hull dimensions of steamboats in the several tonnage classes. This shows that the shallow depth so characteristic of the western steamboat hull was developed gradually over a period of several decades. The trend in this direction got under way in the early twenties, proceeded slowly and without spurts, and was not completed until the decade preceding the Civil War.[42] Clearly the shallow hull and the flat bottom which accompanied it were not the product of an inventive *tour de force* of a single individual.

Still another criterion of steamboat design is the relation of cargo capacity to the size of the vessel, obviously a matter of great importance in the economy of steamboat operations. This is best expressed as a ratio of cargo capacity (in tons by weight) to measured tonnage (cubical content). In the fully developed western steamboat this ratio was ordinarily about 3 to 2 and sometimes reached 2 to 1. Because of their heavy machinery,

[40] Dimensions of the *New Orleans* have been taken from John Melish, *Travels in the United States* (1812), II, 60; those of the remaining vessels from the customs-house enrollment records, Commerce Division. The National Archives.

[41] Ibid. Tonnage figures given here have been adjusted to the old system of measurement in effect prior to the act of 1864.

[42] Tables illustrating this trend and based on enrollment records will be published in a forthcoming work.

shiplike design, and staunch construction,[43] the early steamboats were able to carry much less; the ratio was 2 to 3 or less. The cargo capacities of three of the Fulton steamboats were reported as follows: *Aetna* (360 tons), 200 tons; *Vesuvius* (340 tons), 230 tons; the second *New Orleans* (324 tons), 200 tons. The *Washington*, despite the advantage resulting from the light weight of her machinery, lagged behind all these craft with a reported capacity of but 200 tons although she measured 403 tons.[44]

In the first steamboats, both boilers and engine were placed in the hold, and passenger quarters were located either in the hold or on the main deck. On the fully developed western river boat, boilers and machinery were always on the main deck and the passenger cabin on the deck above. Shreve has been credited with both innovations. "Cynics observed," runs the most fanciful of these attributions, "that with so shallow a hull, there was no place to put the machinery. But when it came time to install the machinery, Shreve gave everybody another shock, by putting it on deck. This seemed the crowning folly. But not so. Since the engines and boilers took up so much of the space on deck, Shreve, nothing daunted, put another one above it, and lo! here was a two decker, the first of the type that was later universal on the Mississippi."[45] Just what Shreve did or did

---

[43] The lightness of draft, which was the prime object of western steamboat design, was a function not only of hull design and proportions but of the weight of the vessel. To reduce weight was as important as to give the hull a broad and flat bottom. It was partly for this reason that the high-pressure engine was favored over the low-pressure type. It was for this reason, too, that the western steamboat came to be built with a lightness of timber, planking, and superstructure that largely justified the epithet of flimsy. There is no evidence that Shreve made any contribution to this important trend. A statement in his report as superintendent of river improvements in 1833 suggests that he regarded it with disfavor. In defense of his method of snag removal, Shreve declared that so far as "good and substantially built" boats were concerned, virtually none had recently been sunk by snags. On the other hand, his report continued, "a great many of the boats now navigating the Mississippi river are light timbered, just sufficient to hold the plank together to bear caulking"—and hence easily sunk by snags. *American State Papers, Military Affairs*, V, 210.

[44] The data on capacity have been taken from the following: *Liberty Hall and Cincinnati Gazette*, September 23, 1816; *Pittsburgh Commonwealth*, February 14, 1816; Felix Flugel, "Pages from a Journal of a Voyage Down the Mississippi," 433; J. H. Morrison, *History of American Steam Navigation* (New York: W. F. Sametz and Company, 1903), 217–219. It is worth noting that on her record trip, New Orleans to Louisville, in the spring of 1817, the *Washington* carried but 155 tons' cargo. *Cincinnati Western Spy*, April 25, 1817.

[45] Hartsough, 45–46.

not do in these as in other matters is uncertain, for the contemporaneous evidence is slim. We know that in the *Washington* the boilers were raised to the main deck but it is equally clear that the engine was placed on the deck below.[46] The *George Washington*, completed in 1825, is customarily described as the first western river boat with an upper cabin, but in this innovation Shreve was anticipated by the builder of at least one boat, the 153-ton *Emerald*, completed in the autumn of 1824.[47]

V

But although Shreve did not invent the flat-bottomed hull, his contributions to the development of the river-boat engine were certainly substantial even if not so great as some writers claim. Indeed these claims have been entirely too sweeping. According to one river historian, who refers to the *Washington* as "the first real steamboat on the western waters," "Shreve contributed three ideas to his 403 ton craft; he placed the machinery and the cabin on the main deck, he used horizontal cylinders with vibrations to the pitmans, and employed a double high-pressure engine. Subsequent marine architecture simply improved on these features."[48] Another writer declares: "The machinery was almost as revolutionary as the architecture of the boat. Not only were the boilers put on the deck but they were placed horizontally instead of vertically. Unlike the upright stationary cylinders that Fulton had used, those on Shreve's boat were placed horizontally and had oscillating pitmans. Furthermore, the *Washington* was the first boat on the western waters to utilize high pressure engines."[49] These claims must now be examined.

The steam engine introduced by Fulton on the western rivers was the low-pressure, condensing engine of the Boulton and Watt type with which his eastern steamboats were equipped. It was efficient and comparatively safe in operation but never attained popularity on the western rivers. Because of its weight, complexity, and cost it was superseded in a few years. The engine that became standard equipment on western river boats was

[46] E. A. Davis and J. C. L. Andreassen, "Form Louisville to New Orleans in 1816. Diary of William Newton Mercer," *Journal of Southern History*, II, 395.

[47] An advertisement of the new steamboat *Emerald* describes her as particularly well suited to passengers, "her cabin being on the *upper deck* [his italics], entirely secured from accidents." *Nashville Whig*, November 1, 1824.

[48] W. J. Petersen, "Steamboats," in *Dictionary of American History*.

[49] Hartsough, 45–46.

a high-pressure, noncondensing, direct-acting, horizontal-cylinder affair with a cam-actuated valve gear. Crude and inefficient from an engineering point of view, it had the practical advantages of being light, compact, powerful, cheap to build, and easy to repair. It was admirably adapted to the conditions of navigation on the western rivers where shallow depth placed a premium on light weight, where swift currents called for great power, where scarcity of skilled labor dictated simplicity of construction and operation, and scarcity of capital favored low cost. Efficiency of operation was a minor factor in a region where wood and coal for fuel were abundant and cheap. In its fully developed form, this engine dominated the western rivers for fully three-score years, and was still in wide use in the first decade of the twentieth century.

The most distinctive feature of the western steamboat engine was the use of steam of high pressure, 50 to 125 pounds compared with the four to eight pounds employed in the Boulton and Watt type. It was this high pressure that made possible a maximum of power with a minimum of weight, the importance of which I have already stressed. The credit for introducing the high-pressure engine in the West belongs not to Shreve, however, but to Oliver Evans. It was Evans who first developed the high-pressure engine in this country, turning out his first engine in 1801, and remaining for some years the only commercial producer of this type of engine in the United States.[50] It was Evans who in private correspondence, in published writings, and in the public press proclaimed widely the superiority of this type of engine to the conventional Boulton and Watt type and who, years before the introduction of steam navigation in the West, urged the special value of his engine for overcoming the difficulties of navigation on the Ohio and Mississippi rivers. The first steam engines used in the trans-Appalachian region were built by Evans. The first steam flour mills in the West went into operation at Pittsburgh and Lexington in 1809, and a third was built at Marietta two years later; and all these mills were equipped with Evans engines. In 1814 Evans listed thirteen engines of his manufacture in operation throughout the West, five of them at Pittsburgh. In 1811–1812, while the first Fulton steamboat was being built at Pittsburgh, Evans was establishing there the first manufactory of steam engines in the West, with his son George in charge of the works. Other men quickly entered the new field, and Pittsburgh became the

---

[50]The data on Oliver Evans that follow have been drawn chiefly from the admirable biography, *Oliver Evans* (1935) by Greville and Dorothy Bathe.

first center of steam-engine and steamboat building west of the Appalachians. A European traveler visiting the rising industrial center in 1816 declared that most of the engines in use were of the Evans type.[51]

Shreve was preceded not only by Oliver Evans but by Daniel French in his appreciation of the importance for western steamboats of an engine combining much power with small weight. French's engine, patented in 1809, was used in the first three steamboats built by the Brownsville group, including the first steamboat to ascend the Mississippi and Ohio rivers, the *Enterprise*. So far as can be determined by the brief references to the French engine, it was direct-acting, dispensing with both beam and flywheel, and it drove a stern paddlewheel. It used steam of high pressure and a contemporaneous account calls it the ideal type for steamboat purposes, since it combined simplicity with compactness and light weight, and had but one tenth the parts of other engines and less than one half as many moving parts.[52]

With Evans and French pioneering in the development of a steam engine that was simple, light, and powerful, what then was the nature of Shreve's contribution? How solidly are the sweeping claims made in his behalf? These claims are based chiefly on two articles about Shreve published in the *Democratic Review*[53] in 1848, a full generation after the episode described. The pertinent passage, referring to the *Washington*, follows:

> Previously, the boiler had always been placed in the hold of the vessel; and under Fulton's patent upright and stationary cylinders used—under French's the vibrating cylinder. Despite the ridicule with which his suggestions were received, he ordered the cylinder to be placed in a horizontal position, and the vibration to be given to the pitman. Fulton and French used a single low-pressure engine; Captain S. built a double, high-pressure engine, (the first used on the western rivers,) with cranks at right angles, and the boilers on the upper deck. Mr. David Prentice had previously

[51] David Thomas, *Travels through the Western Country in the Summer of 1816* (1819), 61.

[52] Communication to the *Cincinnati Western Spy*, February 16, 1816. This was prior to the construction of the *Washington* later in this year.

[53] XXII (1848), 168. Contradicting the assertion here that the French engine was a low-pressure engine are the facts of its light weight, simplicity, and oscillating cylinder, all of which argue against its using low pressure, and the assertion of a rival, George Evans, who declared in a letter to Oliver Evans in 1814 that French used forty pounds pressure and over. Bathe, 217.

employed the *cam* wheel for working the valves to the cylinder; and Capt. Shreve added his great invention of the "cam cut-off," by which three-fifths of the fuel was saved. Most of these improvements, originating with him, have long been in universal use, although their origin has not been generally known. The machinery weighed only one-twentieth as much as the Fulton engine, and was worked with about one-half of the usual amount of fuel. The alterations and improvements by Capt. S. made the engine essentially a new machine; and in the course of a few years, no other model was used west of the Alleghenies.

Contemporaneous references to the machinery of the *Washington*, although very meager, confirm the light weight (given as 9,000 pounds), the great power (100 H.P.), the direct action, the horizontal cylinder, and the location of the boilers on the main deck. Newspaper descriptions of the *Washington* on the occasion of her maiden voyage refer to her engine in the singular.[54] But even the author of the *Democratic Review* articles did not claim for Shreve the cam-actuated valve gear which was so vital a part of this engine. No contemporaneous evidence has come to my attention which supports the same author's assertion that Shreve invented the "cam cut-off," a device by which steam was used expansively by shutting off the flow of steam into the cylinder before the stroke was completed. Major Stephen H. Long, testifying in 1849, declared that he was the one responsible for the cam cut-off, having devised it in 1818 for use on the *Western Engineer*, built at Pittsburgh under his direction for the Yellowstone Expedition of 1819.[55]

The most important feature of the western steamboat engine that appears clearly to stem from Shreve was the horizontal cylinder. The great size and weight of the piston in the low-pressure engine virtually compelled the use of a vertical cylinder in order to avoid excessive friction and wear. With the much smaller cylinder and piston of the high-pressure engine, a horizontal cylinder became a practical possibility. Its use on the

[54] *Liberty Hall and Cincinnati Gazette*, July 1 and September 23, 1816.
[55] *Order of Reference of the Supreme Court in the Wheeling Bridge Case* (Saratoga Springs, 1851), 549–550. Since few western steamboat engines cut off before the completion of five eighths of the stroke, the assertion that the cutoff saved three fifths of the fuel is somewhat exaggerated. Equally to be doubted is the statement that Shreve's engine used but one half the fuel required by other engines. A similar claim was made by Oliver Evans for his high-pressure engine but the verdict of both logic and experience supports the superior fuel economy of the low-pressure engine.

western steamboat was very likely prompted in the first instance by the difficulty of connecting a vertical cylinder engine with a paddlewheel placed at the stern. Since all the steamboats in which the Brownsville group were interested were sternwheelers, French was faced with this problem before Shreve. It is quite possible that he partly anticipated Shreve's solution by giving his oscillating cylinder a position which, within the limits of its motion, approximated a horizontal position. Shreve in turn may have been led to conclude from his experience with the 75-ton *Enterprise* that a French engine of the size necessary to drive the 403-ton *Washington* would prove quite impracticable as a result of the vibrations set up by so large a moving part as the cylinder. By fixing the cylinder in a horizontal position and giving the oscillation to the pitman (connecting rod), this difficulty was overcome. By whatever route arrived at, the adoption of the horizontal cylinder was an important innovation. The horizontal cylinder eliminated the necessity for using large and heavy braces—the clumsy gallows frame of eastern river boats—to give support and rigidity to the engine. Instead, the engine was bolted fast along its entire length to cylinder timbers that were one structurally with the framing of the hull. A maximum of support and stiffness was obtained with the simplest of means, and at a substantial saving of space, weight, and cost. The innovation was to prove of particular importance for the western rivers where the trend toward an increasingly shallow hull would soon have forced some modification of the vertical engine.

For the further claims that Shreve introduced the use of cranks placed at right angles on the stern paddlewheel shaft,[56] that he was the first to drive sidewheels independently by separate engines, thereby greatly increasing the maneuverability of the vessel, and that he was the first to use double flues in boilers, and that he devised a way of supplying the latter through "aft stands," the meager contemporaneous evidence supplies neither confirmation nor contradiction.[57] Writers who have taken the *Democratic Review* articles as their authority for the sweeping claims on behalf of Shreve have failed to note the implications of the anticlimactic remark with which the author concluded his discussion of this subject:

[56] On two-engine boats to avoid stalling on dead center.

[57] Even if Shreve was the first to use two engines and to place the cranks at right angles on sternwheel shafts, he started no revolution. Single-engined boats predominated on the western rivers until well into the forties and sternwheelers played a distinctly minor role to 1850.

"It is possible that, simultaneously with many of his improvements, other persons were, in different parts of the United States or Europe, working out the same results; but if so, it was in this case as in that of Newton and Leibnitz; none the less praise is due to his genius."[58]

Of the five basic features of the western steamboat engine: high-pressure steam, lever-valve gear, direct action, cam cut-off, and horizontal cylinder, only the last stems indisputably from Shreve. That Shreve devised an engine that for its day was very effective is demonstrated by the performance of the *Washington*, although of the precise manner in which this engine was constructed and operated we know very little. It is reasonable to assume that this engine was influential in bringing about the ultimate adoption of the horizontal high-pressure engine, but whether the fully evolved steamboat engine that first comes into view in the late thirties bore anything more than the most general resemblance to that of the *Washington* is a matter for mere conjecture.

VI

With the other bases for Shreve's fame—his role in the overthrow of the Fulton-Livingston monopoly on the lower Mississippi and his activities in river improvement—I need deal but briefly here. Shreve certainly was the leading figure in the attack on the exclusive rights granted the eastern partners by the territory of Orleans. Yet it must be remembered that he played no lone hand in this controversy, and he and his partners received the strong support of public opinion. Moreover, other steamboat owners joined him in refusing to bow to the claims of the monopolists. While tradition portrays Shreve as a plumed knight battling for the right in disregard of personal advantage, it will hardly be denied that substantial financial rewards were directly at stake. The exclusive rights asserted by the eastern firm could hardly have been maintained long in the face of the rising temper of the West, and were, in any event, to be nullified by the Supreme Court's action in *Gibbons* v. *Ogden* within a few years. The spirited opposition led by Shreve no doubt hastened the day when the steamboat would run down the western rivers unvexed to the sea.

Much more important were Shreve's activities as superintendent of western river improvements from 1827 to 1841. In that office he distin-

---

[58] *Democratic Review*, XXII (1848), 168.

guished himself by his success in dealing with the snag problem. He developed a highly effective steam snagboat, he directed the removal of the Raft of the Red River, and he championed a snag-prevention as well as a snag-removal program. Not until the history of western river improvements is written will it be possible to assign Shreve his proper place and importance in this field. In the meantime it is well to remember that snag removal was but one phase of river improvement, although a most pressing one in the early steamboat years, and that credit for progress in this as in the more fundamental aspects of river improvement during these years must be shared with members of the United States Engineer Corps who participated in the planning and execution of these activities. Enthusiasm in the discovery of a new western hero should not lead us to accept some current versions which picture Shreve as solely responsible for the successes achieved while he was superintendent. Those who have accepted the Patent Office decision crediting Shreve with the invention of the snagboat will do well to read the account of its origin by Major Stephen H. Long, Shreve's immediate predecessor and successor as superintendent of western river improvements, later Chief of the Topographical Engineers, and an engineer of distinction.[59] The suggestion of the method for attacking the Raft of the Red River which proved successful originated in the Office of the Chief of Engineers.[60] When Shreve put it into effect, the seemingly Herculean task of clearing a river jammed with trees and logs from shore to shore over a distance of 140 miles assumed in large part the routine character of pulling out and floating off this timber. On the first two days of operations on the Raft, the river was cleared for a distance of five miles and in less than four weeks the Raft was removed over a stretch of forty miles.[61] Not all the going was so easy as this and Shreve's ingenuity and engineering skill counted for much in the final achievement.

Shreve's place as one of the leading mechanics and engineers of the West's industrial infancy is securely established. On the basis of the scanty evidence now available, one can readily support the thesis that he did more for the development of steam navigation on the western rivers than

[59] *See House Doc. 2*, 28th Cong., 1st sess. (Serial 439), 212–213; *see also* selected committee report and attached documents in *House Report 272*, 27th Cong., 3d sess. (Serial 428).

[60] Correspondence between Shreve and General Gratiot, Chief of Engineers, reprinted in J. Fair Harden, "The First Great Western River Captain. A Sketch of the Career of Henry Miller Shreve," *Louisiana Historical Quarterly*, X (1927), 40 ff.

[61] Ibid.

any other one man, including Fulton. Freely to concede this is not to subscribe to the traditional account in which he is pictured as the creator of the American steamboat and the genius "who, by his gifted intellect and untiring perseverance, has opened the arteries of a continent and sent through them the life blood of commerce."[62] From the appearance of the first crude steam vessels on the western waters to the emergence of the fully evolved river steamboat a generation later, we know astonishingly little of the actual course of technological events and we can follow what took place only in its broad outlines. The development of the western steamboat proceeded largely outside the framework of the patent system and in a haze of anonymity. Historians have been too prone to compensate for the wide gaps in our knowledge by playing up the achievements of the few men whose names have come down to us. There is reason to believe that if the returns were all in, the accomplishments of a Fulton, a Shreve, an Evans, or a French would assume a quite modest position beside the collective contribution of scores of master mechanics, ship carpenters, and shop foremen in whose hands the detailed work of construction, adaptation, and innovation largely rested.

[62] *Democratic Review*, XXII (1848), 249.

The "American System" of Manufacturing

ROBERT S. WOODBURY

It is well known that the Industrial Revolution had a profound impact on America; however, not so well known is the fact that America had a lasting influence on industrialism. Innovations like the steam engine and the factory system were transformed as they were assimilated within American technological traditions. The "American System" of manufacturing was an outstanding example. From its origins in arms manufacturing to the assembly line of Henry Ford, there is an unbroken continuity—a tradition that is, therefore, one of America's most distinctive contributions to technology. But, as with the western steamboat, the study of its origins has been obscured by legend and the heroic theory of invention. In the following selection, Robert S. Woodbury, a professor of history of technology at the Massachusetts Institute of Technology, attempts to set the record straight. Professor Woodbury is the author of a series of monographs on the history of machine tools.

In some legends the story is such that from its very nature we can never establish its truth or falsity; in others patient historical work—usually external to the legend—can ascertain whether the events acually happened or not. The legend of Eli Whitney's part in interchangeable manufacture is, however, unique in that the clues and even much of the evidence for its refutation are part of the legend as customarily recited. It is also unique in that the legend is not merely a popular one nor even a story given "au-

Robert S. Woodbury, "The Legend of Eli Whitney and Interchangeable Parts," *Technology and Culture*, 1 (Summer 1960), 235–253. Copyright © 1960 by the Society for the History of Technology. Reprinted without footnotes by permission of the author and the publisher, The University of Chicago Press.

thority" by inclusion in conventional textbooks. This legend has been retold at least twice with all the paraphernalia of historical scholarship—footnotes, elaborate bibliography, discussion of the sources, and even use of archival material. But in both cases we find the same failure to evaluate the evidence critically, to follow leads to other sources, and to question basic presuppositions. These same faults extend back to the origins of the legend.

Poking back into the beginnings of this legend, one finds evidence to show that it was at least partially created consciously by its hero and uncritically accepted by most of his contemporaries. The *editio princeps* of the legend is equally uncritical; in fact it is frankly an *apologia pro vita sua*. In his *Memoir of Eli Whitney*, Denison Olmstead gives us most of the elements of the legend and claims to have based his account upon conversations with those who knew Whitney, as well as upon examination of his correspondence and Miller's. Yet Blake writing in 1887 said: ". . . there have not been wanting persons who have endeavored to take from Mr. Whitney the credit of originating the uniformity system and making it a great practical success at the beginning of this century, thus leading in the van of progress of the mechanical arts, and laying the foundations for the enormous industry development of the nineteenth century." Evidently some of his contemporaries were not taken in by Whitney's claims, but the scholars have not asked either who these other inventors were or what their contributions may have been. Let us examine the principal parts of this legend in some detail.

I

THE CONTRACT

Whitney's contract of June 14, 1798 to manufacture arms for the Federal Government is the focus of a number of elements of our legend. His motives in this undertaking have been interpreted as those of a prudent businessman doing his patriotic duty and as those of a genius anxious to put into execution a new scheme of manufacture for the good of his country in a time of crisis. His actual motives were quite different.

In 1798 Miller and Whitney had lost all their suits to obtain their cotton gin patent rights in the courts of the South. What little legal merits these decisions had, stemmed from a defect in the Patent Law of 1793;

clearly nothing further could be done until Congress corrected this defect. The efforts of Whitney and others did not finally result in a new patent law until 1800. The intervening years could be seen as a lull in the affairs of Miller and Whitney. But Whitney could hardly look forward to any relaxation, for their financial affairs were in desperate straits. Every source of credit had been exhausted by both partners. Certainly Whitney himself was on the verge of a nervous breakdown. Although some have tried to find in this situation a frustrated love for Catherine Greene, a more careful reading of his letter of October 7, 1797 to Miller indicates rather that Whitney's high hopes of financial security, respected position, and prestige have not only come crashing to the ground, but the disgrace of bankruptcy is staring him in the face. All that winter of 1797–98 Whitney brooded alone, half-heartedly carrying on the affairs of Miller and Whitney. He shut himself off from all his old friends and even unjustly accused his partner and friend Miller.

Whitney needed a new opportunity—any opportunity. But, more important, he needed credit—credit to save Miller and Whitney from bankruptcy, credit to enable him to fight for his rightful profit and for his good name lost in the cotton gin suits. When he heard that the Congress was "about making some appropriations for procuring Arms etc. for the U.S.," here was a heaven-sent opportunity. This would at least keep his manufactory going until he could get his cotton gin rights. The opportunity was so great and Whitney's situation so desperate that he was willing to promise "ten or Fifteen Thousand stand of Arms," a fantastic proposal! Whitney even promised to begin delivering "in a short time" and he "will come forward to Philadelphia immediately. . . ." New hope for a desperate man!

Why was such a rash proposal not rejected at once by such prudent men as President Adams and Timothy Pickering, the Secretary of War? The failure of the Pinckney mission had caused public concern, and French privateers were rumored to be off the coast. Even Washington was called out of retirement to head the armed forces. On the 4th of May 1798 Congress voted $800,000 for the purchase of cannon and small arms. When on the 24th of May Whitney arrived at the seat of government the plum was not only ripe and juicy but begging to be picked. Public sentiment was aroused, and the highest officials must do something—and that right promptly. Both sides could not close the contract quickly enough. Only the Purveyor of Public Supplies, Tench Coxe, seems to have kept his head —"I have my doubts about this matter and suspect that Mr. Whitney cannot perform as to time."

It is not necessary to see "influence" at work here, though it is true that Whitney did have a number of Yale graduates who could help him. Much less is there any evidence that the generous terms of Whitney's contract grew out of the feeling that he had been shabbily used in his cotton gin suits. But he clearly had a personal friendship with Jefferson arising out of the patent for the cotton gin.

Certain features of the contract deserve closer examination. The legend makes much of the fact that the actual document was wholly handwritten. It says that all the other contractors of this time received printed contracts, and that there was therefore something special about Whitney's contract. Unfortunately an examination of the actual contracts, including Whitney's, in the National Archives shows that this was by no means the only handwritten contract—there were others, such as that of Owen Evans of Providence, Penn. The fact is that several of the early contracts were handwritten; the later contracts, mostly signed in September, were printed forms. These other contracts, printed or handwritten, were all identical in wording and provisions with Whitney's, except in the terms of the last paragraph. There *was* something special about Whitney's contract—it contained a paragraph six not included in any of the others. It was this paragraph that was crucial for Whitney. Having quickly sized up the situation in which the high officials found themselves, the shrewd Whitney saw his chance, consulted Baldwin as to the form the contract should take, and at one stroke solved all his immediate problems. This paragraph reads:

> 6th. Five thousand dollars shall be advanced to the party of the second part on closing this contract, and on producing satisfactory evidence to the party of the first, that the said advance has been expended in making preparatory arrangements for the manufacture of arms, Five Thousand dollars more shall be advanced. No further advances shall be demanded until One thousand stands of Arms are ready for delivery, at which time the further sum of Five thousand dollars, shall be advanced. After the delivery of One thousand stands of arms, and the payment of the third advance as aforesaid, further advances shall be made at the discretion of the Secretary of the Treasury in proportion to the progress made in executing this contract. It is however understood and agreed by and between the parties to this instrument, that from time to time, whenever the party of the second part shall have the second thousand ready for

delivery he shall be intitled to full payment for the same, so with respect to each and every Thousand until he shall have delivered the said Ten thousand stands.

Here was credit at last! Here was financial standing which assured further credit! Five thousand dollars at once, and five thousand more on terms which could be easily fulfilled by using his cotton gin laborers and machines. And assured payment for each thousand of arms upon delivery —to a total of $134,000. Little wonder that Whitney wrote to his friend Stebbins: "Bankruptcy and ruin were constantly staring me in the face and disappointment trip'd me up every step I attempted to take, I was miserable ... Loaded with a Debt of 3 or 4000 Dollars, without resources, and without any business that would ever furnish me a support, I knew not which way to turn ... By this contract I obtained some thousands of Dollars in advance which saved me from ruin."

No wonder that in his eagerness to read paragraph six of the contract, Whitney evidently skimmed rapidly over the incredible terms of paragraph one. Whitney had contracted to deliver 4000 stands of arms by September 30, 1799, and 6000 more by September 30, 1800. Four thousand stands of arms in 15 months, from a factory yet to be built, and made by laborers as yet untrained, and by methods as yet unknown! And 6000 more in the following year! In his desperation Whitney had thrown all caution to the winds. He was no experienced manufacturer, as his deliveries of the relatively simple cotton gin indicate. He was aware that he knew nothing of arms making. And a prudent man would have expected at least some of the setbacks with which he fills his later letters to Wolcott, together with requests for further credit, contrary to the provisions of the contract. In short, despite his vague claims of new methods and what could be done by "Machinery moved by water," Whitney had only the vaguest idea of how he would actually fulfill the contract. He was not able to deliver even the first 500 muskets until September 26, 1801, and the contract was not actually completed until January 23, 1809. Further, the records of the Springfield Armory, now in the National Archives, show that even during the period 1815 to 1825, when his plant was fully established, Whitney never delivered muskets at the rate promised in his contract of 1798.

Not only have these facts been forgotten in estimating Whitney's motives in the contract, but also in attempting a proper evaluation of his troubles with Samuel Dexter, who had replaced Wolcott as Secretary of the Treasury. We are asked to see Dexter as a villain abusing our hero with "malice" by demanding that he perform in accordance with his contract. The other contractors of 1798 had in many cases failed to fulfill their contracts, and some of them had even gone into bankruptcy as a result of their efforts to manufacture arms for the federal government. But Springfield Armory records show that some of them had performed as contracted, a few on time and even more eventually. Yet these men had all ventured into arms making by financing themselves privately. There are no records of their writing the long apologetic letters full of troubles, promises, and requests for further advances, which characterize Whitney's correspondence from 1798 on. Nor had they been given the numerous informal extensions of time with which Wolcott, strongly under the influence of Hamilton's theories of the importance of manufactures, had favored Whitney and which culminated in a formal modification of the contract just before Wolcott left office. Whitney had been given more consideration than any other contractor.

But one might give at least a moment to the position of Dexter. He was a government official sworn to carry out the law and to protect the interests of the government. Whitney had been given every chance and had not performed. Some of the other contractors had. Even had Dexter seen the ultimate interests of the government in this matter in the broad terms that Wolcott and Jefferson did, he had no authority to make the extremely loose interpretation of Whitney's contract that Wolcott had. Actually Wolcott had left office partially as a result of other similar easy exercise of the discretions permitted his high office. Dexter did not deserve such blame.

And had Whitney, for his part, acted in good faith since 1801? We can leave out of our discussion the troubles Whitney so fully related in his numerous lengthy letters to Wolcott. They were real enough, even if recounted in rather unmanly fashion, but they were all of the sort, magnitude, and frequency which a prudent man would expect in an undertaking of this sort. And one could argue that if Whitney had been carelessly optimistic in what he had promised in the contract of 1798, so had the responsible government officials, who had also been warned by the Purveyor of Public Supplies to expect delay in delivery. However, despite Whitney's claims of the exhausting efforts and attention he had devoted to his arms manufacturing, the facts prove otherwise.

It is true that the lull in the affairs of Miller and Whitney from 1798 to 1801, plus the credit advanced him by the federal government, did enable

Whitney to devote most of his time in these years to make a beginning on fulfilling his arms contract, and by September of 1801 he did deliver the first 500 muskets. But from this initial delivery until 1807 there is no twelve-month period during which he delivered over 1000 muskets. During this same period he had been given advances from the Treasury such that he was constantly in debt to the United States. In fact, when Whitney finally completed delivery in January 1809 he received a payment of only $2450 as final settlement of the total contract of $134,000. Only on this date was his account up to date. The Whitney account in the Springfield Armory records also shows that in 1806 Whitney delivered 1500 muskets, in 1807 he delivered 2000, and in 1808 and the first few days of 1809 he delivered 1500. What is the explanation of these facts?

I do not wish to imply that Whitney was misrepresenting his troubles in his letters to Wolcott, Dexter, and Dearborn; but he most certainly was not telling the whole story, as his other correspondence clearly shows. In April 1800 the Congress revised the patent law which had been the legal means of defeating Whitney's claims to his cotton gin rights. Under the new law Miller at once started suit against the principal offenders. The "lull" was over. But it became increasingly evident that justice would not be done Miller and Whitney in Georgia under any law. On September 4, 1801, Miller wrote to Whitney of a new possibility—their patent rights were to be purchased by the state legislatures. Here was a greater reward than Whitney could have dreamed of! Miller needs Whitney's help and his "contacts." Whitney cannot wait and by November 22, 1801 is dating a letter to Stebbins: "Virginia Nineteen Miles North of the Northern line of North Carolina." He sees a chance that Miller and Whitney may get $100,000 from South Carolina alone for his rights—here was freedom from debt, assured financial security, and a credit reputation of the best. Better still, he will have the fame and prestige of a name officially cleared and full credit for his invention. Is it any wonder that a man of Whitney's ambitions and self-interest rushed off to Columbia and left the troubles and problems of arms manufacturing behind? Whitney had slaved and scrimped to get through Yale that he might become respected and financially secure. Now fortune beckoned, and the arms contract could wait.

From the fall of 1801 until Judge Johnson's decision of December 19, 1806 the ups and downs of cotton gin affairs were certainly far more important in Whitney's mind than the manufacture of arms. This can be definitely established simply by noting the places where Whitney's and others' letters show him to have been in this five-year period. The contents of his correspondence clearly establish a similar conclusion, as does the mere volume of the lawsuits in which Miller and Whitney were engaged. In the final settlement of the partnership in 1818 Whitney was allowed $11,000 for the expenses of six journeys South on these lawsuits. Certainly he was seldom attending full time to the arms manufactory at New Haven.

In short, from 1801 to 1806 Whitney not only failed to fulfill the contract, he regularly substituted long letters of excuse for honest effort to carry out his obligation, while he chased the richer prize of the rewards he expected from the cotton gin. In the light of these facts Dexter can hardly be blamed for his actions, and Jefferson's intervention seems hardly to have been in the interests of the government, whatever effect it may have had upon the future of American industry.

II
MANUFACTURE BY THE UNIFORMITY PRINCIPLE

The shortage of skilled artisans in the formative years of the American republic has been so often repeated as the source of Yankee mechanical ingenuity that it is now taken as axiomatic, without careful examination of the actual numbers as adequate for the needs of the day. This same axiom has served to "explain" Whitney's use of manufacture by interchangeable parts. In fact, Whitney so explains it himself. But let us look at the facts. The Springfield Armory was opened in 1794, and its payroll records from the beginning are to be found in the National Archives. By 1802 the Armory had 76 skilled armorers employed, and by 1814 it had 225. Although the figures for Harper's Ferry have not been preserved and we know it to have been substantially smaller than Springfield, we would be safe in assuming that Harper's Ferry had at least half this number of armorers. This total is impressive and seems hardly to indicate a scarcity of skilled armorers. In addition, we have the records of deliveries by other private contractors of arms to the Springfield Armory. During the whole period which concerns us, either the Springfield Armory, or Asa Waters of Sutton, Mass., or Lemuel Pomeroy of Pittsfield, Mass., delivered at least as many arms in each year as did Whitney. In fact Springfield manufactured 16,120 in the six years from 1795–1801, a much more impressive record than Whitney's 10,000 in ten and a half years. Both started from nothing. Leaving out of account the deliveries of the smaller manufacturers, Springfield, Waters, and Pomeroy certainly had an ample supply of armorers—or are we to believe that they too had the principle of interchangeable parts which Whitney claimed was unique in his establishment at New Haven?

But where did Whitney get his ideas for manufacture of arms on this new principle? He always claimed that it was his and his alone, and so the legend says, despite strong evidence to the contrary. There can be no doubt that prior to Whitney other men had actually used the principle of manufacture by interchangeable parts. In the 1720s Christopher Polhem, in Sweden, was manufacturing gears for clocks by using machinery and precision measurement to ensure interchangeability. But there is no evidence that Whitney or anyone in the United States knew of Polhem's work, though it could have influenced Blanc in Europe.

The work of Blanc [sic] was clearly known to Thomas Jefferson; in fact our legend always includes a recital of his letter to John Jay in 1785 describing Blanc's work, and Jefferson's letter to Monroe of November 14, 1801, in which he points out that by 1801 Whitney had not developed the method as far as Blanc had in 1788. But the most amazing thing about the Whitney legend is the failure of scholars to follow up this clear lead to answer two questions of first importance: (1) Who was Blanc and what did he do? (2) Did Blanc's work have any influence upon Whitney?

The sources on Blanc are not only easily available, but are very detailed on his methods and results, for much of his work was done in French government arsenals and created controversies which were the subject of several official investigations and reports. Even a cursory examination of these sources would indicate that Whitney was far from being the first to introduce the principle of interchangeable parts in the manufacture of small arms. It is also quite clear that Blanc had carried the technique much further than we have any evidence for Whitney's doing. Furthermore, Blanc's *Mémoire* of 1790 shows a profound understanding of the nature and probable effects of interchangeable manufacture, of which Whitney had only the barest inkling. Whitney's goal was only a system to use unskilled labor to increase output and reduce cost; whatever interchangeability he achieved was only a by-product of his method.

Blanc had problems to meet that Whitney never did. An entrenched officialdom and a threatened craft labor in long established government arsenals, together with the eclipse of the nascent industrial revolution in France under the Revolution and Napoleon, prevented a final fruition and spread of Blanc's ideas and methods in France and on the Continent.

But did the spark fly from Blanc to Whitney? A careful search in the correspondence of Whitney, Jefferson, Monroe, Jay, and Stiles indicates only that there were at least several paths by which it may very well have passed, of which the most likely is through Whitney's numerous conversations with Jefferson. But the only positive evidence seems to be a letter from Wolcott to Whitney dated 9 October 1798 in which he encloses ". . . a pamphlet on manufacture of arms . . . inform me freely and candidly whether the performance appears to you calculated to afford instructions to the workmen in this country. . . ." Whitney replied on 17 October 1798 that it was "misleading." Can this have been a copy of the report by le Roy on Blanc's *Mémoire* of 1790? We know that Jefferson was a regular subscriber to French publications, including the *Encyclopédie* as issued in parts. Did Jefferson's interest in Blanc in 1785 lead to receipt of this publication of the Académie des Sciences? And did he pass it on to Wolcott? One more bit of evidence remains. Writing to Wolcott on December 24, 1800, Whitney regrets that Wolcott does not have the leisure to examine "my whole plan and manner of executing the different branches of the work . . . to . . . compare them with the modes practiced in this and other Countries." Was Whitney actually familiar with the methods in use abroad, and if so, through what means? One must also admit that the language of many of Whitney's letters describing the merits of his methods are strangely reminiscent of Blanc's words. Yet we have nothing conclusive.

We must also ask whether Whitney's contemporaries in America may have influenced him, in particular the work being done at the Springfield Armory. It is most significant that after signing the contract in June 1798 Whitney had gone to Springfield to see their methods and to talk with the superintendent. And we have Whitney's letter in the summer of 1799 in which he had originally written "I might bribe workmen from Springfield to come to make me such tools as they have there." It is clear that Whitney was prepared to copy at least some of the methods already in use at Springfield. What were these machines? Unfortunately a fire in 1801 destroyed many of the records of the Springfield Armory, and the question cannot be answered fully. But we do have one official report that gives clear indications that special machinery was in use by, at the latest, 1799:

". . . the artificers were employed for some time on the buildings, instead of on the manufactory, and in making the necessary pieces of machinery and tools. . . ." [If we take into account the difficulties of opening a new establishment, such as] "unsuccessful attempts in the proper construction of the machinery," [we should be satisfied with the present cost of muskets]. The report also uses such expressions as "The works now being complete, and labor-saving machines operating to great advantage . . . ," and ". . . improvements in the machinery and system for carrying on the manufactory."

That these improvements in the machines were effective is shown by the fact that in the month of September 1798 the Armory produced 80 muskets, but the following September 1799 it produced 442 muskets. This was accomplished with the same number of workers on the payroll. The report goes on to state that it had previously required 21 man-days to produce a musket; with the improved machinery only 9 man-days were needed. This at a time when Whitney had not yet delivered a single musket!

The later correspondence between Whitney and Roswell Lee, then superintendent at Springfield, although lacking technical details, strongly suggests that, contrary to Whitney's claims, at least a simultaneous development was going on. And there are patents, contracts, and accounts of Simeon North that strongly suggest that he, too, was using interchangeability in making his pistols as early as 1807.

John Hall begun work on his rifle designed to be made by interchangeable parts and on machinery to manufacture it prior to his patent of 1811 and was installing his methods and machines in the Harper's Ferry Armory by 1817. In 1827 Hall petitioned the government to give him adequate recompense for his contributions. This resulted in a series of commissions and investigations to establish the facts, by which he was finally compensated in 1840. The reports of these boards are matters of public record. The most significant for our purposes is one of 1827—two years after Whitneys' death.

> In making this examination our attention was directed, in the first place, for several days, to viewing the operations of the numerous machines which were exhibited to us by the inventor, John H. Hall. Captain Hall has formed and adopted a system of manufacture of small arms, *entirely novel*, and which, no doubt, may be attended with the most beneficial results to the country, especially if carried into effect on a large scale.

His machines for this purpose . . . are used for cutting iron and steel, and for excuting woodwork . . . and *differ materially from any other machines we have ever seen in any other establishment . . . By no other process known to us (and we have seen most, if not all, that are in use in the United States) could arms be made so exactly alike as to interchange . . .*") [Italics mine.]

This report was signed by James Carrington and Luther Sage, who had been government arms inspectors for years and were thoroughly familiar with the methods in use at Springfield and in the manufactories of the private contractors, including Whitney's. That Whitney himself thought Hall's work new is shown by the fact that he made the long trip to Harper's Ferry to see the "new system being adapted there."

A later report indicates that the machinery was especially desirable for it could manufacture "all other species of arm *identically*." This later report also shows that the machinery had been in use since at least as early as 1819 at Harper's Ferry: "At Harper's Ferry, and at Springfield, this machinery is believed to be *exclusively* used; and the money expended upon it, and upon the tools at the former armory from 1819 to 1834, both inclusive, was within a fraction of $150,000." The commission stated that since Hall was employed as an armorer at Harper's Ferry after 1819 he deserved no compensation in addition to his regular pay for improvements made after that date. But it recommended that he be compensated for the work he did from 1811 to 1819.

All this can hardly be said to justify our legend's categorical statement, "In every way Hall profitted by Whitney's work."

III
MANUFACTURE BY MACHINERY

We have thus far taken the term "manufacture by interchangeable parts" to have a clear meaning, based upon Blanc's, Whitney's, and Hall's dramatic demonstrations in assembling arms out of parts taken at random. This is a concept based upon characteristics of the product. It of course raises the question of how closely the parts must fit to be interchangeable. The usual answer is that the tolerances allowed must be sufficiently small for the product to work as designed and no more, since closer tolerances will merely increase cost. But this is rather vague. A more significant concept of interchangeable parts results from an examination of the actual methods by which such parts are produced. In this sense modern interchangeable parts require these elements: (1) precision machine tools, (2) precision gauges or other instruments of measurement, (3) uniformly accepted measurement standards, and (4) certain techniques of mechanical drawing. We do not, of course, expect Whitney to have all these elements, but we can estimate the contribution he may have made by comparing his work to them.

In what sense were the Whitney firearms interchangeable? A test of a number of known Whitney arms in at least one collection proved that they were *not* interchangeable in all their parts! In fact, in some respects they are not even approximately interchangeable! The answer to this paradox is to be found partly in the actual means of establishment of standards for their manufacture. Each of the contractors of 1798 (and the later contractors as well) was given two or three samples of the Charleville model of 1763, and his contract specified that these were to be followed exactly. This method meant that at best the output of one plant would be interchangeable, but the muskets of a given contractor would not necessarily be interchangeable with those of the other contractors. In short, our third and fourth elements of interchangeable parts—uniform standards of measurement, and working from adequately dimensioned drawings—were absent. In fact, they were not to appear for two more generations.

However, the first steps in this direction were to be taken by John Hall. Writing to Congress February 21, 1840, he says: "And so in manufacturing a limb of a gun so as to conform to a model, by shifting the points, as convenience requires, from which the work is *gauged* and executed; the slight variations are added to each other in the progress of the work, so as to prevent uniformity. The course which I adopted to avoid this difficulty was, *to perform and gauge every operation on a limb from one point*, called a bearing, so that the variation in any operation could only be the single one from that point." [Italics mine.]

What about our second element—use of gauges? Polhem had used these, and so had Blanc. There is clear evidence that gauges were being used at the Springfield Armory by 1801. Hall certainly had used them extensively before he went to Harper's Ferry in 1817, but there is not the slightest evidence that Whitney ever did.

A number of visitors went through the Whitneyville plant in Whitney's lifetime. All were properly amazed, but none wrote an account which tells us what Whitney's actual methods were, except that there were "mould" and "machines." By putting bits of information together, the "moulds" can be interpreted as what would today be called die forging; Blanc had clearly used this method. But "moulds' may also refer to filing jigs. The legend makes much of: (1) the numerous references by Whitney and others to his "machines," (2) the machine tools listed in the inventory made by Baldwin of Whitney's estate, and (3) Whitney's supposed invention of the milling machine.

Let us examine each of these in detail. First, we may ask what did the term "machine" mean in Whitney's day? It most certainly did not mean what it does today. It included a trip hammer and a water wheel, but it also meant almost any kind of device. What machines did Whitney actually employ? In this connection we have the letter of ten-year-old Philos Blake, Whitney's nephew, written after his visit in September 1801: "Thare is a drilling machine and a boureing machine to bour berels and a screw machine and too great large buildings, one nother shop and a stocking shop to stocking guns in, a blacksmith shop and a trip hammer shop and five hundred guns done." This is the only first-hand evidence we have of Whitney's machines at this time. Yet an official inspection of the Springfield Armory in January 1801 says the following: ". . . the number of Files required at the Factory being so great, some Water Machinery is now preparing which will diminish the demand of this expensive article." Even more advanced machinery was used in the national armories by 1817.

Timothy Dwight, one of Whitney's visitors prior to 1823, says: "Machinery moved by water . . . is employed for hammering, cutting, turning, perforating, grinding, polishing, etc." But by this time we have clear evidence that such machinery was in use at both Springfield and Harper's Ferry.

The list of machine tools in the inventory of Whitney's estate is detailed and tells us much about the tools he had in use at the time of his death in 1825, but lists nothing not already in use at Springfield and Harper's Ferry. In fact, the large number of files listed as on hand would suggest that for much of his work Whitney used only a filing jig or fixture to guide a hand operated file as his principal means of producing uniform parts for the locks of his muskets. But Polhem had done this two generations earlier.

The Whitney papers at Yale also include a number of drawings, none of them dated, signed, or even identifiable as definitely made by Whitney;

these drawings are quite possibly those of Whitney's nephews, for Benjamin Silliman, writing in 1832, says: "The manufactory has advanced, in these respects [machinery], since it has been superintended by Mr. Whitney's nephews, the Messrs. Blake, and to them it is indebted for some valuable improvements." They had been in charge for about five years before Whitney's death.

The legend includes one specific machine—the milling machine discovered in 1912 by Professor Joseph W. Roe of Yale, and now in the collection of the New Haven Colony Historical Society. It was identified by Eli Whitney's grandson of the same name as having been made by his grandfather and as the first one ever made. His authority for this identification was that he remembered having seen it as a boy and having been told this story by workmen in the old Whitneyville plant. Roe dated this machine as of 1818 merely because of a statement in the *Encyclopaedia Britannica* that "the first milling machine was made in a gun manufactory in 1818." All this hardly seems adequate evidence. The first reference we have to the use of milling by Whitney is in his letter to Calhoun of March 20, 1823. But by 1818 we have clear evidence that milling was in common use in both national armories, and by at least Robert Johnson and Lemuel Pomeroy of the private contractors.

In short, we really know practically nothing of what Whitney actually had in his manufactory at Mill Rock; what little we do know of was clearly not an innovation; and we have good evidence to show that all that Whitney claimed as his own contribution was at least independently innovated by others, particularly in the national armories. Whitney's claims of originality seem to have been the exact opposite of the truth. Certainly no one is justified in stating that Whitneyville was the site of "the birth of the machine tool industry," much less the birthplace, even in America, of manufacture by means of interchangeable parts.

IV
"THE BIRTH OF AMERICAN TECHNOLOGY"

There can be no doubt that what became by the 1850s widely known abroad as the "American system of manufacturing" had its origin in this first quarter of the nineteenth century and that its principal features were developed in the northeastern section of the United States. The American system included mass manufacture, by power-driven machinery, by machinery especially designed to serve its particular purpose, and by the use of the principle of interchangeable parts.

The legend says that all this stems from Eli Whitney. We have seen enough to indicate that we actually know very little of what he really did; hence there is no clear beginning from which we can tell what later developed from Whitney's work. It is also clear that other men were working along these very lines in the manufacture of arms at the same time.

The legend also claims that from Whitney stemmed the application of the American system of manufacture to many light metalworking industries—Colt and his revolver, Jerome's clocks, Waltham watches, Yale's locks, Singer's sewing machines, and so on. Even if we knew exactly what Whitney did, there is little evidence to support this application of the-great-man-in-history hypothesis. About all that can be said is that further applications of interchangeable parts would *logically* seem to follow from Whitney's broad *claims*. But this is not the same as proof that Whitney actually had methods similar to those of later innovators, much less that they really did derive their ideas and methods from his. Certainly many other men contributed as much or more than Whitney, and evidence for their work can be found, far more convincing than Whitney's boasting claims. The legend says, for example, that the influence of Whitney was the basis of the Colt Armory methods of manufacture. In fact, it was E. K. Root who was the technical genius behind the manufacture of the Colt revolver, and his work stems directly from that of John Hall at Harper's Ferry. Whitney's influence on the manufacture of clocks, watches, and sewing machines is equally open to question.

We know so little of Whitney's actual methods of manufacture that his contribution to interchangeable parts is difficult to assess. What little we do know indicates, if anything, that Whitney was on the wrong track anyhow. John Hall's methods can be fairly clearly established, at least sufficiently for us to be sure that modern interchangeable manufacture derives far more from his inventive genius at Harper's Ferry than from Eli Whitney's manufactory at Mill Rock. Actually one is led to find the origins of the "American system of manufacturing" in the culmination of a number of economic, social, and technical forces brought to bear on manufacture by several men of genius, of whom Whitney can only be said to have been *perhaps* one.

V
CONCLUSION

This analysis of the Legend of Eli Whitney and Interchangeable Parts raises more questions than it answers. We have by no means arrived at the truth about the legend, much less about the advent of manufacture by interchangeable parts. However, I hope it is clear that the whole question needs re-examination—a more critical analysis of presuppositions and of the evidence which is known, and a more careful search for other sources.

But why not let this nice convenient legend go on? Were it Whitney alone that concerns us, that might be well enough. But the issue is larger than that. The history of our industrial growth is of first importance to the understanding of our American heritage. That industrial development cannot be properly understood without careful consideration of its technological basis. Therefore the true story of the "Birth of American Technology" is of prime concern to us. We should make certain that the baby is perfect and legitimate.

The Direction of Technology

BRUCE SINCLAIR

If technology is not directed by a few legendary figures, then who is responsible for the developments that have taken place? Since technology is, on one level, a synthesis of knowledge, scientific societies provide an obvious source of direction. The Franklin Institute is an exemplary case. Founded to provide supplementary education for apprentices, it evolved into a professional organization that could speak for the technological community. By the mid-nineteenth century, technology—aided by such institutions—began to move from artisanship toward professionalism. In this selection, Bruce Sinclair shows how the Franklin Institute, aided by the federal government, promoted a standard American system of screw threads. He also demonstrates how seemingly minor issues may illuminate important social realities—the standardization of screw threads helped advance the American system of mass production. Bruce Sinclair is a professor of the history of technology at the University of Toronto and editor of *Early Research at the Franklin Institute* (Philadelphia, 1966). Currently, Professor Sinclair is writing a history of the Franklin Institute.

At a time when the superb drama of exploiting a new continent filled the minds of most Americans, nineteenth-century technical arguments about the shape and number of threads on a screw often have a remote, somewhat comic quality—a bit like Jonathan Swift's mock epic struggle between the Little Enders and the Big Enders. But the issue was not a matter of satirical trivia. Industrial development on a national scale demanded that nuts and bolts of the same diameter be interchangeable. Interchangeability, in turn, required that manufacturers conform to a standard

Bruce Sinclair, "At the Turn of a Screw: William Sellers, the Franklin Institute, and a Standard American Thread," *Technology and Culture*, 10 (January 1969), 20–34. Copyright © 1969, the Society for the History of Technology. Reprinted without footnotes by permission of the author and the publisher, The University of Chicago Press.

system which fixed the contour of screw threads and established for each diameter the number of threads per inch. For America, in 1860, no such standard existed. Since that era also marked the emergence of this country into the arena of international industrial competition, the search for a standard cast reflections which illuminate such related considerations as a national style of engineering, American industrial practice, and the role of government in technological change.

The most prevalent system—where a system was used at all—was that which had first been proposed in 1841 by England's Sir Joseph Whitworth. Whitworth's standard was a synthesis of the best English practice and answered the general case well enough so that it was widely employed. American usage was disuniform, however, varying from manufacturer to manufacturer and from locality to locality. Some firms developed systems to fit the particular requirements of their own processes. Others purposely used special threads to prevent outside repairs on their own machinery, in the same vein as the Erie Railroad's ill-fated use of wide-gauge track. By the 1860s, the appalling lack of national uniformity clearly called for reform. "If there is any one thing in the transactions of the machine shop more incomprehensible than another," the editor of *Scientific American* claimed in 1863, "it is the want of some settled size or number for screw threads." To eliminate the anarchy, the magazine called for some agreement among the country's principal manufacturers or, failing that, governmental action. Whatever the standard, Whitworth or any other, the adoption of some uniform national system of screw threads was of vital necessity.

But the development of a standard in the first instance posed certain difficulties. As Whitworth had pointed out, a system depended on compromise, not on theory or experimentation. There was simply no way in which all the factors involved in screw-thread design could be stated as a precise rule, applicable to all cases. Where principle provided no final answer, practice determined the issue. That opened the door to non-technical factors. A further difficulty in America was that no single agency seemed capable of dealing with the problem. Editorial injunctions notwithstanding, the federal government had no ready means either to develop a system or to regulate its usage, not even to speak of an inclination to do so. Nor did any professional engineering society exist to propose a standard and advance its adoption. It was in this apparent vacuum that William Sellers presented a paper before a meeting of the Franklin Institute at Philadelphia in April 1864 outlining a uniform system for American screw threads.

In his paper, Sellers addressed himself immediately to the central question: Why should there be yet another system of threads? Why should not Americans adopt the Whitworth standard and by its consistent usage rationalize current practice? The English system was the result of several years of study by an outstanding mechanician, who had carefully analyzed the three main factors of screw-thread design—pitch, or the number of threads per inch, thread depth, and thread form. To arrive at his system of pitches, Whitworth had collected bolt samples from all of the principal manufacturers in England, averaged their characteristics, and developed a standard table for pitch and diameter. The form of his thread, which comprised flat sides at an angle of fifty-five degrees, with rounded tops and bottoms, was also arrived at by the same averaging technique. It was a compromise, but on the basis of the best English screw-thread practice. Any alternative scheme, Sellers suggested, should clearly "demonstrate its practicability and its superiority."

The central difference between Whitworth's system and the one proposed by Sellers was in the form of thread. Sellers leveled three objections to the English form. First, the fifty-five degree angle was difficult to gauge with consistent accuracy. Second, in ordinary practice the rounded tops and bottoms of English threads did not fit the corresponding bottoms and tops of nuts, and the thread's wearing surface was therefore reduced only to its sides. The point was not that a fit was impossible, but that in normal usage it was difficult to achieve contact with that form of thread. Finally, Sellers objected to Whitworth's thread form on the grounds that it was more complicated, and therefore more costly to manufacture. According to Sellers, obtaining the rounded top required three kinds of cutters and two lathes to perform what with our practice requires but one cutter and one lathe." The core of Sellers' opposition to the Whitworth standard was that it was complicated, expensive, and difficult to produce with consistent accuracy.

In place of the Whitworth form, Sellers proposed a contour then already in use in several Philadelphia machine shops, including his own. The thread took the shape of an equilateral triangle, sides inclined at an angle of sixty degrees, with the top and bottom flattened one-eighth of the thread depth (see Fig. 1). It had, Sellers claimed, precisely those virtues which were defects in the Whitworth form; it could be made with greater

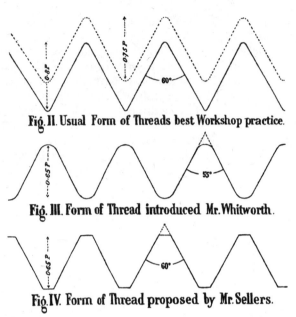

Fig. 1. Figure II above shows normal screw-thread practice when no particular standard was employed. Figures III and IV outline the characteristics of the Whitworth and Sellers threads. From Robert Briggs, "A Uniform System of Screw Threads," *Journal of the Franklin Institute*, LXXIX (February 1865), 124.

ease and less machinery, and could be verified with greater accuracy. The table of pitches which Sellers proposed varied only slightly from the Whitworth standard, but was notable primarily because those variations allowed for the use of a relatively simple formula to calculate pitch for any diameter. To his thread form and table of pitches, Sellers also proposed a uniform range of sizes for bolt heads and nuts and a standard thread gauge, thus offering for consideration a complete and standardized system of screw threads, nuts, and bolts.

Since screw threads were not a matter of objective precision, as Whitworth had pointed out, the determination of a system obviously lent itself to bias. Even particular elements of a screw thread reveal the possibilities. For example, assuming a generally V-shaped contour, pitch is the most significant factor in uniformity. It is important that, for any given diameter, the number of threads per inch is the same. The close similarity between the pitches of Sellers' system and Whitworth's suggests that on this issue there was little to separate American from English practice. But the form of Sellers' screw thread highlights the differences in American engineering and industrial attitudes. Nowhere was that bias more clearly revealed than in the subsequent competition between the two thread systems.

One of the keynotes in the battle was sounded almost immediately. William Bement, head of the Philadelphia machinery firm of Bement and Dougherty and chairman of the Franklin Institute committee to investigate and report on Sellers' system, set the tone when he advised the editor of *Scientific American* that the committee would be pleased to have the opinions of "all good *practical* mechanics." The emphasis of the word was Bement's. The magazine reiterated the idea, congratulating American mechanics that their interests had not fallen "into the toils of schemers and theorists who would have confused instead of making the subject plain and practical." What was meant by "practical" was made clear in the Institute's report. The Sellers thread form was based on an angle "more readily obtained than any other." It was a thread contour which any ordinary workman could make with accuracy. The committee's conclusion was that a uniform American thread system should embody a shape which would enable "any intelligent mechanic to construct it without any special tools." There was nothing theoretical or abstruse about it. It sounded democratic and it saved money.

All subsequent consideration of the Whitworth and Sellers thread systems reiterated the basic difference in thread form. The U.S. Navy's Bureau of Steam Engineering conducted a fairly extensive analysis of screw-thread systems in 1868, with a view toward the adoption of a standard for the Navy. "We find but two *systems* in general use," the Board noted, "one known as Whitworth's, deduced many years ago from the general practice of English mechanics, and the other known as that of Mr. William Sellers, deduced more recently from the practice of American mechanics." In regard to pitch ranges, strength, and durability, the Board of Naval Engineers found no appreciable difference between the two systems. According to their report, the form of the Sellers thread was one of the major factors which led the Board to recommend his system. The Whitworth thread required "such skill on the part of the workmen" that uniformity was out of the question. Conversely, the Sellers thread enjoyed the "very important advantage of ease of production."

In England, uniform screw threads depended on a high level of craft skill. And to Englishmen, such skill seemed a perfectly appropriate foundation. As one editorialist succinctly put it: "With good tools and good men to use them, there is no difficulty whatever in producing the Whitworth cross-section of thread." Where labor costs were lower and craft traditions stronger, the system worked. But it was for precisely those reasons that Whitworth's system was rejected here. In America, interchangeability depended on reducing a job to the point where an ordinary workman could produce the desired uniformity. It was not Whitworth's intention to conserve labor. The fifty-five degree angle of his thread was an arithmetic mean; it was mathematically indifferent to production costs. American engineering practice had labor costs as one of its basic considerations, and Sellers' system reflected the difference. The genius of his standard, however, was that it was not only easier and cheaper to construct, but easier as well to maintain with accuracy.

Because his thread perpetuated his name, Sellers is probably more widely known for his system of threads than for any other facet of his career. But its acceptance depended in part on his position in American industrial and engineering circles. William Sellers was Philadelphia's leading machine-tool builder when that city was the country's preeminent tool-building center. "Probably no one," according to Joseph Roe, "has had a greater influence on machine tools in America than William Sellers." At a time when it seemed to some that the screw-thread problem lacked a solution simply for want of any leadership, Sellers provided it. Although he had just celebrated his fortieth birthday when he presented his paper on a uniform thread system, Sellers had already established a reputation as one of Philadelphia's leading machinists. He came to his trade early. It was one of the characteristics of the city's industrial elite to go directly to business, not college. After a basic education at a private school maintained by his family, Sellers was apprenticed to his uncle's machine shop in Wilmington, Delaware. Following his apprenticeship, he superintended the family-connected machine shop of Fairbanks, Bancroft and Company in Providence, Rhode Island, for three years and then, in 1848, returned to Philadelphia where he established a partnership with Bancroft in the manufacture of machine tools and mill gearing. So successful was the firm, a contemporary observer remarked, that within less than a decade it was the city's leading machine tool company.

Sellers was frequently styled the "American Whitworth," and there is no reason to believe he shunned the comparison. In fact, there is a remarkable parallelism in the accomplishments for which the two men were remembered. Both conducted large and successful industrial enterprises. Both advocated high standards of precision in machine-shop practice, and to that end both produced machines whose form depended on function rather than on artificial architectural embellishment. But there were also important differences between the two. Whitworth's passion was accuracy. For him, "a *true plane* and *power of measurement*" were the outstanding elements of mechanical engineering, and the major thrust of his career was to upgrade English practice along those lines. Sellers was by no means unconcerned with precision, but it would be fair to say that his work showed greater originality. In the sense that Whitworth's interest was with the maintenance and extension of precision workmanship, he was a conservator; by comparison, Sellers was an innovator.

From the beginning, Sellers' machine tools were dominated by the same thinking which characterized his screw-thread system. The catalogues of his firm explicitly stated that position:

> It has been said that good workmen can do good work with poor tools. Skill and ingenuity may indeed accomplish great results, but the problem of the day is not only how to secure more good workmen, but how to enable such workmen as are at our command to do good work, and how to enable the many really skillful mechanics to accomplish more and better work than heretofore; in other words, the attention of engineers is constantly directed to so perfect machine tools as to utilize unskilled labor.

Sellers' machines embodied that view. His 1857 bolt machine marked the advent of commercially interchangeable nuts and bolts. The gear-cutting machine, which he showed at the Paris Exhibition of 1867, was one of the earliest to be automatically operated. And he produced an increasing number of special-purpose lathes on the principle that they would allow "less skillful workmen" to produce more work of better quality.

Sellers was also a grand-scale industrialist. He organized the Edge Moor Iron Company in 1868, which supplied all the structural iron work for

Philadelphia's Centennial Exhibition and all structural materials except the cables for the Brooklyn Bridge. Significantly, at Edge Moor bridge-making was reduced to a standardized manufacturing process. In 1873 Sellers also became president of Midvale Steel Company, where he supported Frederick W. Taylor's long series of experiments in metal cutting. By middle age Sellers was already the dean of Philadelphia's machine industry; at his death, he was recognized as "the greatest tool builder of his day and generation." But it was Whitworth himself who bestowed the highest accolade when he described Sellers as "the greatest mechanical engineer of the world."

By talent and reputation, Sellers was probably better qualified than any other American machinist to effect a substantial change in the country's screw-thread practice. Proposing a system, however, was only half of the battle. The utility of a standard lay equally in its widespread acceptance. Whitworth had recognized the dual nature of the problem. He had presented his own system at meetings of the Institution of Civil Engineers, and he was aware of the importance of engineering opinion for the system's adoption. Not only was it difficult to achieve agreement on a system, inconvenience in the change-over and the inertia of traditional usage would also inhibit the spread of a standard. These were "obvious" reasons, Whitworth claimed, why engineers and their societies should push for the adoption of a uniform system. Sellers was faced with similar impediments, but he was closely connected with major elements of private industry and was president of the only technical society in the country which enjoyed a national reputation—the Franklin Institute of the State of Pennsylvania. Therein lay the means for adoption.

A general society in the pre-professional era of American science and technology, the Franklin Institute was founded in 1824 as one of the earliest American mechanics' institutes, with the aim of educating young apprentices in science and its applications. In the decades which followed, a series of technical studies projected the organization into national prominence. Its experiments into the motive force of water provided an intensive analysis of hydraulic power. The Institute's investigation into the causes of steam boiler explosions was the most exhaustive and detailed study conducted in this country and remained so for another half-century. The organization acted as technical advisor to the state of Pennsylvania on a standard of weights and measures and to the city of Philadelphia on the best method of paving streets. It participated in the investigation into the

U.S.S. *Princeton* disaster and, along with the National Academy of Sciences, co-operated in a Navy Department study of steam expansion in 1863. These activities gave the Franklin Institute a reputation for informed judgment, and that prominence was a significant factor in the adoption of Sellers' system.

Equally important, there was a solid link between the Institute and Philadelphia's leading machinery firms. The committee which was appointed to consider the subject of screw threads was mainly composed of representatives from those firms. Chairman of the committee was William Bement, of Bement and Dougherty, a firm second only to William Sellers and Company in the city's machine-tool hierarchy. Baldwin's Locomotive Works had two men on the committee. The Southwark Foundry and Merrick and Sons, sprawling industrial firms established by Samuel V. Merrick, one of the founders of the Franklin Institute, were also represented, as were the firms of Morris, Towne and Company, the Pencoyd Iron Works, and of course, William Sellers and Company. The Institute's Board of Managers bore the same stamp. When William Sellers was elected president of the organization in 1864, most of the same firms were represented on the Board. Most of them were also using Sellers' system in their own practices. In their combined interests, these men and their firms wielded enormous influence in the American mechanical community. Promoters of rival thread systems may have felt cause to complain, as one did, that the committee simply confirmed Sellers' conclusions rather than conduct a thorough investigation of the whole problem.

From the very beginning, the Institute zealously pushed for widespread adoption of the Sellers, or, as it was soon also called, the Franklin Institute system. A special committee was appointed to consider the subject raised by Sellers' paper, to the end that the Institute would recommend a system "for the general adoption of American engineers." At the same time, copies of Sellers' paper were sent to other mechanics' institutes, "with a view to promote the introduction into general use of the system advocated." When the special committee made its report, it was unanimously in favor of Sellers' plan. The Institute then moved to secure general adoption by American engineers. Formal resolutions were sent to the quartermaster general of the Army, chief of the Navy's Bureau of Steam Engineering, the chiefs of ordnance of both the Army and Navy, the chiefs of the engineer and military railroad corps, and to the superintendents and master mechanics of railroad companies. The resolutions called for the adoption of the system "by requiring all builders under any new contracts to conform to the proportions recommended." Similar resolutions were sent to all mechanics' institutes and to the major machine shops of the country.

Governmental approval was the most obvious and direct means of obtaining conformity, and the Institute's resolutions were heavily weighted in that direction. The Navy was the first to respond. In March 1868, the secretary of the Navy authorized the Bureau of Steam Engineering to recommend a standard screw-thread system for the service. Benjamin Isherwood, chief of the Bureau, appointed Theodore Zeller, chief engineer at the Philadelphia Navy Yard, to head a board of naval engineers for that purpose. After visiting the principal machinery establishments in the country and after formal meetings in Philadelphia, the Board in its report of May 1868 "unhesitatingly" recommended adoption of the Sellers system. By direction of the secretary of the Navy, it was immediately authorized as the Navy standard.

Since the same Philadelphia machinery interests which backed Sellers' system were then involved in a nasty dispute with Isherwood and Zeller, the Bureau's quick accord is perhaps surprising. The quarrel involved used machinery which Zeller had purchased from a New York dealer for use at the Philadelphia Navy Yard. Local machinery manufacturers resented the loss of business to an outsider and raised serious questions about the transaction. The issue led to a congressional investigation which reinforced an already bubbling intraservice squabble, and Isherwood ultimately lost his job. But the Bureau's action suggests the importance of the thread system's identification with the Franklin Institute. The *Journal of the Franklin Institute* had long been a favorite vehicle of professional expression for ambitious young Navy engineers. And when both the Navy and the National Academy of Sciences declined to publish Isherwood's *Experimental Researches in Steam-Engineering*, the Franklin Institute did so. It is also true that the Navy advocated Sellers' system in the belief that it would probably become the established practice in private industry. As the Board stated in its report, "We were naturally desirous to select a system which, while meeting all the essential requirements of a system, would be most likely to be generally acquiesced in and adopted." In other words, the Navy's concern was with Navy practice; it had no concept of leading a movement for the Sellers standard. Private interests were important only to the extent that they had already reached a consensus in favor of the system.

The Board's conclusions effectively threw the burden of standards enforcement back onto private shoulders, and in the last analysis private interests were the most significant element in the adoption of the Sellers system. Two factors shaped the response of industry. First, the standard reinforced technical advances in machine tools; it meant maximum utilization of self-acting and automatic machinery. The laborious process of thread cutting by hand had long since given way to the use of self-acting lathes, which employed change gears to obtain different pitches. And after 1860, machine shops had automatic bolt cutters at their disposal, further to speed and standardize the thread-cutting process. Improvements in machinery made it all the easier to manufacture a standard screw. The adoption of a system rationalized those technical advances.

A standard system also had particular benefits for certain segments of industry, and it was from those areas that the most immediate and effective support came. Interchangeability was critical on railroads, for example. It was obviously important that locomotives and rolling stock be manufactured according to standards, since repairs were frequently required at points distant from a central shop. Steam engines and the equipment of heavy industry fell into the same category, and acceptance came first from these sources. By 1868, Sellers' system was used in most Philadelphia firms and had also been widely adopted by New England machine shops and by several manufacturers of machinery in the mid-Atlantic states. The Pennsylvania Railroad took up the system in 1869. It was adopted by the Master Car-Builders' Association and the Master Mechanics' Association (both railroad organizations) shortly afterward and, within the next decade, by most of the country's railroads. Standard gauging tools for the system, developed by Pratt and Whitney and by Browne and Sharp, only hastened the process of adoption.

But discussion and controversy did not end. General American acceptance of the system raised the debate to an international level, where it became entangled in the metric system controversy and in engineering competition between nations. For example, when German engineers proposed the adoption of the Sellers thread form for use in a metric system to replace the Whitworth standard, English technical response was sharp. The effort, according to one editor, was an eccentricity on the part of "a handful of obscure scientific men." It ran counter to the enlightened practice of thirty years and would bring ruin to Germany's export trade. Furthermore, it was claimed that after a brief period of experimentation in America, the Sellers thread contour proved to be a conclusive failure, and like errant children come home, machine makers in the United States were returning to the Whitworth pattern.

Faced with increasing engineering competition from America and from a Continent potentially united by the metric system, English technical opinion was confused and divided. The Verein Deutscher Ingenieure (Society of German Engineers), also perplexed by the notion that there might not be an American standard thread system, addressed a communication of inquiry to the Franklin Institute, with particular reference to thread form. The Institute's secretary, Dr. William H. Wahl, adopted a novel form of response. He directed a circular letter to the officers of the major American railroads and to a number of manufacturing firms, asking whether they employed the Sellers standard and, if so, how successfully the thread form was maintained. The replies, together with a counter-blast from *Railway Gazette*, were all forwarded to the German society.

It is difficult to say which offended the Americans most—English condescension, aspersions on their engineering practice, or technical inaccuracy. *Railway Gazette*'s editor retorted hotly: "This is not the first time that facts and figures as to American practice have been 'evolved from the inner consciousness' of writers across the water, to suit the occasion, but it is not often that such a complete perversion and reversal of the facts is given currency in a journal of standing." According to George Bond, of Pratt and Whitney, 90 percent of the orders they received for taps and dies were for the Sellers system. Furthermore, Bond claimed that experimentation and two decades of experience had proved the demonstrable superiority of its thread form. There was an American standard, Dr. Wahl informed German engineers. It was called the Sellers, or Franklin Institute, standard and was used "throughout the United States, to the exclusion of any other." The exchange reflected, once again, the importance of nontechnical elements in the problem of screw-thread standardization. International industrial competition and engineering nationalism were the issues.

The establishment of professional engineering societies in America did not change the equation. Debate continued along much the same lines, though the forum was shifted to specialized groups. At the first annual meeting of the American Society of Mechanical Engineers, George R. Stetson, of the Morse Twist Drill and Machine Company in New Bedford, pointed out that one of the functions of the new society should be to push the acceptance of the Franklin Institute standard so that "American machinery, by is uniformity in approved design and construction, may be entitled to the most favorable consideration in foreign as well as home markets, and the field of our usefulness and profits correspondingly increased." In fact, the A.S.M.E. failed to reach agreement on any standard, primarily on the principle that such questions were best left for solution in the marketplace. As Monte Calvert has noted, it was not that the society was opposed to standards, but that it felt they should be determined by businessmen and business methods.

Trade associations, business firms, and engineers working individually—not government—established national screw-thread practice in nineteenth-century America. The Navy's adoption of Sellers' system was important insofar as it set a precedent and created a general climate of accord. Writers on the subject always established the government's acceptance of the system to suggest that it enjoyed a posture of authority. But there was no reality of governmental enforcement. Implementation was always a job for private hands. In England, the subject was a matter of debate in the House of Commons. In France, it became settled governmental policy. But in America, private interests, more than any other, determined the outcome. It was by the efforts of entrepreneurially minded technical groups like the Franklin Institute and the "untiring devotion" of organizations such as the Master Car-Builders Associaiton that uniformity was advanced in America.

To many, it was right that such questions as screw-thread systems should be resolved by practical men rather than by politicians. Coleman Sellers spoke to that point of view when he remarked: "The government of France has always been in the habit of interfering with the private affairs of people." Conversely, he noted, the American concept of governmental function was "that it should do and enforce justice, and that Liberty in all things innocent, is the birthright of the citizen." In time, that attitude changed; simple answers no longer fit the problem. As specialized societies examined the technical requirements of their own groups more closely, the weaknesses of a general system of screw threads became more apparent (see Fig. 2). That called, in turn, for a review of the whole issue, suggesting some governmental action. With the establishment of the Bureau or Standards in 1901, the government also had a means of responding to the problem. Subsequent efforts involved combinations of government and technical societies.

Nothing better illustrates the nineteenth-century attitude than George

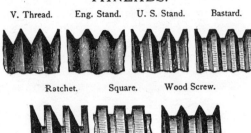

Fig. 2. This illustration shows the most commonly used screw threads of the nineteenth century. Whitworth's thread form is shown as the English standard; Sellers' thread is labeled as the U.S. standard. The V thread and bastard thread are variants from Whitworth and Sellers models. The remainder are special-purpose threads. From Cornell University Machine Shop, *A Chapter of Advice on Bolts and Nuts with Cards of Reference, Prepared for the Students in the Machine Shop of the Sibley College* (Ithaca, N.Y., 1875), pp. 9–10.

Bond's frequent comment that, among screw-thread systems, it was "survival of the fittest." William Sellers' uniform system was fittest in the sense that it answered contemporary engineering requirements. Sellers understood the economic as well as the technical aspects of the problem and appreciated their interrelationship as a prerequisite for adoption. The Franklin Institute's indorsement and proselytizing activities created a favorable environment for survival. At a time when no similar association existed to perform the function, the organization gave Sellers an institutional framework for his system, providing a platform, a mechanism for its advancement, and an aura of objectivity. The man and the institution were well mated, and it was the combination which produced the Sellers, or Franklin Institute, standard of uniform screw threads.

The Ideology of Technology

HUGO A. MEIER

The most important agency for directing technology is the community. Where technological change has been seen as a danger, it has been subject to social control, as by craft guilds and government in the Middle Ages. The rapid advance of modern mechanical technology in America since 1800 was possible because it received support from the community. In the following selection, Hugo A. Meier points out how some of this support was won. Spokesmen for the new technology argued that it would advance American democracy and strengthen American institutions. We may interpret the result as a powerful ideological defense of unrestrained technological change. Few contemporary Americans, however, would be as confident of the beneficent effects of technology as were their nineteenth-century countrymen. Hugo A. Meier is a professor of history at Pennsylvania State University and is author of a forthcoming study of technology and American society.

The human relationships of technology have become in recent years of increasing interest to the historian as well as to the social scientist. American historians, no longer satisfied that a listing of inventions adequately explains the impact of technology on society, are now trying to place technology in its true perspective within the history of ideas. They have shown an increasing awareness that the concept of technology is linked to long-familiar social ideas, such as nationalism and sectionalism, and to the idea of democracy. In the half century following the Jeffersonian revolution, one may find a rapid if often still crude progress in technology ac-

Hugo A. Meier, "Technology and Democracy, 1800–1860," *Mississippi Valley Historical Review*, 43 (March 1957), 618–640. Copyright © 1957 by the Organization of American Historians. Reprinted without footnotes by permission.

companying and contributing to the major evolution of several great social ideas.

One of these ideas, democracy, became consciously and elaborately associated with American progress in the applied sciences. Evident in the early years of the republic, this association came to emphasize the special role of technology in providing the physical means of achieving democratic objectives of political, social, and economic equality, and it placed science and invention at the very center of the age's faith in progress. At the same time there arose an apprehension of the dangers of an exaggerated materialism as a social consequence of the emphasis on technology in a democratic form of society where, as observers from abroad remarked, concern with physical comforts and conveniences already tended to dominate intellectual, moral, and aesthetic interests.

Technology, to be sure, was not a term current in 1800. "The useful arts" was perhaps the phrase most commonly used in popular writings about applied science, and it was not until Jacob Bigelow, the first Rumford professor "of the application of science to the art of living" at Harvard College, published his *Elements of Technology* in 1829 that the new term entered into popular usage, although Bigelow had long used it in his lectures. Nor was "democracy," with its widespread connotation of mob rule and indiscriminate social and economic leveling, a term acceptably hygienic among many respectable people at the beginning of the century. But the ideas of expanded political liberty and social and economic opportunity, of equality in class status, of the welfare of the many as opposed to the special privileges of the few, had long been taking root in America. Since the days of revolutionary turmoil such "democratic" concepts were gradually interpenetrating the more familiar ideas of "republicanism." It was the general republican notion of civil liberty combined with popular enlightenment, however, which first encouraged American engineers and inventors, as well as many others, to link the fortunes of technology with those of the new American political and social system.

Dramatic instances of this attempt to associate applied science with the fate of republicanism are to be found in Robert Fulton's earnest efforts to win governmental support for his grandiose canal schemes and novel inventions for war. His "creative canals," Fulton argued in 1807, would unite the new republic economically and socially, greatly encouraging and securing the blessings of domestic republican institutions. He promised, too, that his "torpedoes" and his "plunging boat" or submarine would be a means of defending and extending republican institutions. "Every order of things," Fulton generalized, "which has a tendency to remove oppression and meliorate the condition of man directing his ambition to useful industry, is, in effect, republican."

While individuals like Fulton sought to demonstrate that republicanism could be greatly assisted by technology, other citizens were developing the notion that progress in technology itself was powerfully dependent on the notions of equality and liberty inherent in republicanism. Even Europeans, the *American Journal of Science* reported in 1822, were beginning to see that the rapid American progress in discovery and the great number of citizens of all classes participating in it were the product of republican freedoms unknown abroad. Similarly, one of the early American technological journals commented that the cause of such progress in America was the "degree of civil liberty which leaves the human mind untrammeled to avail itself of its own strength." And how else might one explain—in the 1850s—the triumphs of the yacht *America* on the high seas and of the McCormick reapers in the international exhibitions? "When there is free labor upon a free soil, a free head and a free heart to direct, and a free hand to do, we need have no fear of the result," was the confident conclusion.

American technology and American civil and intellectual liberty, in the eyes of representative spokesmen, were marching hand in hand in these years of rapid national growth. Could the Old World—tradition-bound, shackled in mind and spirit by monarchical institutions—profit from the example? America, at least, could show the way. Levi Woodbury, whose career in state and federal politics closely involved him in the ferment of Jacksonian democracy, thought such leadership a genuine responsibility. America, said Woodbury, must no longer seek only to equal the commercial and technological achievements of older powers. She had a new mission: "It is to demonstrate to the old world, by deeds no less than reasoning, that our new theory of private rights and public duties is conducive to progress in everything useful."

Indeed, it was in this stress on usefulness that technology proved to be a catalyst, blending the ideas of republicanism with the rising democratic spirit in the early national period. Americans always had been pragmatically inclined; on occasion they felt obliged to apologize for their pragmatism, offering their explanation in terms of the nature of the environment and of their society. Suspecting that an English man of science might question the devotion of a scholar to the improvement of so humble an instrument as the plow, Thomas Jefferson carefully explained to the president of the Board of Agriculture in London that "the combination of a *theory* which may satisfy the learned, with a *practice* intelligible to the most unlettered labourer, will be acceptable to the two most useful classes of society." In much the same manner, the loquacious Federalist, Thomas Greene Fessenden, excused the utilitarian content of his new *Register of Arts* in 1808 on the ground that the mechanics of splitting logs and rooting out stumps was necessarily of greater interest to Americans than was mere scientific theory. And with considerable editorial frankness the *Useful Cabinet* cautioned the public not to search in its pages for "refined philosophical speculations."

Even in these early years, however, the American stress on usefulness had to answer the skeptical criticism of those who felt utility a shallow criterion—a continuing controversy among intellectuals of this half century. It was objected, for example, that the exceedingly pragmatic approach of American science was a threat to progress in science itself. In answer to the *Useful Cabinet* one troubled reader cautioned that America, encouraged by necessity and enterprise, sought so avidly after the luxuries of life that not only had not "the stream of science become her favorite beverage," but ignorance of basic principles was resulting in fruitless experiments. John Redman Coxe introduced his *Emporium of Arts and Sciences* in 1812 as something of an antidote to mere utilitarianism by promising to devote much of its space to advances in the science of Europe "from whence we must reasonably look for information for many years to come." Dr. Thomas Cooper, English-born scientist and Jeffersonian pamphleteer, seconded such sentiments by reminding the magazine's readers that Americans were not yet so enlightened a people that they need pay no attention to mathematical and physical science.

But the utilitarian bias of American science was not to be gainsaid, and in its stress on the everyday needs of the average citizen it demanded that science descend from its ivory tower and serve the people. Thomas Jefferson, no idler in matters theoretical, stressed the essentially democratic function of American science in his correspondence with Thomas Cooper in 1812. "You know the just esteem which attached itself to Dr. Franklin's science, because he always endeavored to direct it to something useful in private life," said Jefferson. "The chemists have not been attentive enough to this. I have wished to see their science applied to domestic objects, to malting, for instance, brewing, making cider, to fermentation and distillation generally, to the making of bread, butter, cheese, soap, to the incubation of eggs, etc. And I am happy to observe some of these titles in the syllabus of your lecture. I hope you will make the chemistry of these subjects intelligible to our good housewives." Nor was Jefferson's advice a mere sop to empiricism. He was tremendously proud of the practical and largely self-contained economy of Monticello. Similar advice to direct science to common needs was given in 1816 by Jacob Bigelow when he assumed his new post as professor of applied sciences at Harvard. In his inaugural address Bigelow urged that inventive genius be concentrated on the needs of the immediate neighborhood, aiming always at improving "the facilities of subsistence, and the welfare of those among whom we live." Unlike the laborious speculations of the Germans which, he said, probably added little to the real stock of knowledge, "the researches of most of our ingenious men have had utility for their object."

With this kind of approval, one scarcely wonders at the practical trend of American technology by the time of Jackson's presidency, and particularly at the way in which emphasis on applied science had come to influence educational practice. Education in the young democracy reflected its technological emphasis. The roots for practical education extended into colonial times, as shown in Franklin's 1743 "Proposal for Promoting Useful Knowledge among the British Plantations in America," foreshadowing the beginning of the American Philosophical Society. Before the Revolution, too, Franklin was recommending the inclusion of courses in practical science to the trustees of the Philadelphia Academy. Soon after the turn of the century Jefferson was proposing a similar emphasis on "technical philosophy" in his own scheme of education, with the special interests of mariners, carpenters, brewers, clockmakers, machinists, and similar artisans in mind.

In the years after the War of 1812 educational curricula were clearly reflecting the interest in technological learning. The versatile Jacob Bigelow made Bostonians increasingly aware of the subject of "technology" in his popular lectures on applied science as first incumbent of the Count Rumford chair at Harvard, a chair established to teach "the utility of the physical and mathematical sciences for the improvement of the useful arts, and for the extension of the industry, prosperity, happiness and well-being of society." Thomas Cooper, too, was among those interested in applying theory to the needs of a democratic society. He carried his practical approach to chemistry with him from Dickinson College to the University of Pennsylvania in 1816, where he was appointed to the chair of applied chemistry and mineralogy. In Philadelphia, too, the Franklin Institute opened its doors in 1824 to the serious-minded artisan, and in Troy, New York, in 1825, the school which was to become Rensselaer Polytechnic Institute welcomed its first students who, under the direction of Amos Eaton, sought to achieve "the application of science to the common purposes of life." Elsewhere by 1835 "mechanics' institutes" and several reputable college programs in branches of technology had been introduced.

There was no escaping the exceedingly pragmatic view of science in the United States by 1830. And by that time, too, the trend toward identifying this utilitarian outlook with the "American way," democracy, encouraged speculation as to the cause of the relationship. The discerning French visitor Alexis de Tocqueville was one of the contemporary observers of Jacksonian democracy who pointed to the surprisingly close relationship between the social system of the United States and its technological emphases. Explaining, in his *Democracy in America*, "Why the Americans Are More Addicted to Practical Than to Theoretical Science," Tocqueville pointed out that the concept of equality encouraged a taste for the tangible and the real, and a contempt for tradition and forms. "Those who cultivate the sciences amongst a democratic people," Tocqueville argued, "are always afraid of losing their way in visionary speculation. They mistrust systems; they adhere closely to facts and the study of facts with their own senses." Furthermore, the extreme social mobility in America was fertile soil for progress in technology, because democratic peoples were ambitious, never satisfied with their status, and—above all—were always free to change it. "To minds thus predisposed," Tocqueville explained, "every new method which leads by a shorter road to wealth, every machine which spares labor, every instrument which diminishes the cost of production, every discovery which facilitates pleasures or augments them, seems to be the grandest effort of the human intellect. . . . You may be sure that the more a nation is democratic, enlightened, and free, the greater will be the number of these interested promoters of scientific genius, and the more will discoveries immediately applicable to productive industry confer gain, fame, and even power on their authors."

If the utilitarian propensities of a democratic environment proved favorable to technological development as Tocqueville described it in the

1830s, no less real was the relationship of that technology after 1800 to the democratic goal of improving the lot of the common run of man. Indeed, the very utilitarian base of technology made it an instrument especially useful for furthering the ideal of "the greatest good for the greatest number." Since to the American "good" was in many ways becoming synonymous with "goods" (as Tocqueville so frequently emphasized), the goal of mass production became a conscious obligation imposed upon science and invention. In America, science through technology must multiply the resources of human enjoyment and universalize their availability. An example of belief in "the greatest good for the greatest number" is concealed in Jefferson's praise in 1815 of the invention of a machine to apply steam to the "small and more numerous calls of life," for he believed that "a smaller agent, applicable to our daily concerns, is infinitely more valuable than the greatest which can be used only for great objects. For these interest the few alone, the former the many." This point of view grew more popular as the century advanced. "The true glory and excellency of Science consists in its aptitude to meliorate the condition of man," commented an observer in 1827, while Levi Woodbury in the 1840s concluded that the attempt to benefit the many in practical ways had led to some of the greatest efforts and achievements of American science. He pointed, by way of illustration, to Franklin's invention of the lightning rod and his improvement in design of the common stove.

How uniquely American seemed this concern for the welfare of the many rather than the luxury of the few was epitomized in London in 1851 when the nations of the world displayed the products of their genius and industry at the Crystal Palace. The humble reapers were America's proudest contribution, and to Edward Riddle, who reported to the commissioner of patents on the exhibition, they epitomized an eloquent contrast between democracy and despotism. The fruits of democracy were visible, too, in the stacks of machine-made water pails, pegged boots and shoes, garden tools, bell telegraphs, spring chairs, cooking ranges, and hot air furnaces. "The Russian exhibition," Riddle reported in contrast, "was a proof of the wealth, power, enterprise, and intelligence of Nicholas; that of the United States an evidence of the ingenuity, industry, and capacity of a free and educated people. The one was the ukase of an emperor to the notabilities of Europe; the other the epistle of a people to the workingmen of the world. . . . We showed the results of pure democracy upon the industry of men."

When the burgeoning republic held its own Crystal Palace Exhibition in New York in 1853, "the results of pure democracy" again dominated the American departments. Horace Greeley, the *Tribune's* energetic publisher, summed up the creative forces underlying a half century's technological progress in glowing terms. "We have *universalized* all the beautiful and glorious results of industry and skill," Greeley said. "We have made them a common possession of the people.... We have democratized the means and appliances of a higher life." Like Jefferson's chemistry that brewed and baked, technology was, indeed, in the service of the people.

The influence present and future of the new machine production on the social status of American labor was another aspect of the relationship between technology and democracy which interested many Americans, and troubled some of them, after 1800. Did the machines, the product of America's technological skills and interests, mean a decline in independence and dignity for the laborers who operated them? Higher wages, the opportunity to change jobs, and the comparative newness of machine industry in America no doubt helped to keep the dignity of machine labor in Jacksonian America above European levels. But even the carefully shepherded girls in the early Lowell textile mills felt obliged to make extensive apologies for their kind of employment until the influx of cheap immigrant labor settled the question of status for them. That marvelous little literary effort, the *Lowell Offering*, occasionally hinted at the difficulties. "You must be 'school-ma-am' while you are here, for factory girls are nothing thought of in this place," Miss Matilda was reminded when visiting a former mill companion now returned to her home town. But Matilda was faithful to her Lowell training: "I could not long endure such bondage, and resolved to return where I could enjoy a dearly-loved freedom," she reported to *Offering* readers.

Matilda's sentiments were matched by other contemporary defenders of factory labor. The Reverend Stephen H. Tyng, for example, found the lot of the Lowell ladies a pleasant one—their machinery beautiful and varied, their workrooms clean, their moral purity as yet unspotted. "But above all," he observed, "the universal and mental advantages which are freely provided for every class of operatives, so that the girl from the factory may become, without difficulty or remark, the teacher in the seminary, and the lady of the parlor, are all facts of American peculiarity and great American honor." Other interested supporters of the rising factory system and of technological advance spoke warmly of the beneficent influence of machinery on the status of labor. They pointed out, for example, to those who condemned the bustling factory world of mid-century that the power loom and spinning jenny had freed thousands of pallid women from the poisonous atmosphere of the small shop, while even in agriculture machinery was now emancipating men from its dullest, most tedious and unremunerative employment.

If the machine and the factory system found ready defense against the charges that they degraded the dignity of labor in a democracy, no less an effort was being made to elevated the social status of the men whose talents and labors were creating the machines. A self-conscious technical journalism which sprang into being shortly after the turn of the century helped this new cause by bolstering the ego of the expanding class of inventors and engineers. Technology was demonstrated to be a high and honorable calling, and its practitioners were strengthened in their growing professional consciousness against snobbish exponents of the greater dignity of the law, of politics, and of religion. The inventor and the engineer, it seemed, were at last coming into rightful esteem.

The belief that such a change in status was well deserved found one of its strongest supporters in Thomas Ewbank, whose administration of the Patent Office from 1849 to 1852 was energetic and creative. Ewbank, who had emigrated from England in 1819, could look back from mid-century upon an active career as manufacturer, inventor, and technical writer. His experiences and especially his work as commissioner of patents stimulated his interest in the social relationships of technology, and his pen was active in that cause. For technologists Ewbank had high regard, and he insisted that "The time is not distant when such men, instead of being deemed, as under the old regime, virtual serfs, will exert an influence in society commensurate with their contribution to its welfare." Political fame, he argued, was ephemeral but "no fame is more certain or more durable than that which arises from useful inventions." To Ewbank, the inventor and the engineer were leaders of men.

Though the dignity of engineer, inventor, or machine laborer was granted, commentators on technology in this era confronted another question quite as serious. In a democracy, economic opportunity supposedly was more freely open to all classes than was true in less fortunate lands. Had invention and the introduction of machinery on an extensive scale harmonized with this happy doctrine? From industrial Europe filtered tales of strikes and anti-machinery riots, of miserable, underpaid men, women, and children laboring in dismal factories. What, indeed, were the economic prospects of American labor in a machine-minded democracy?

Pioneer promoters of American industrialism—men like Alexander Hamilton and Tench Coxe—had pointed out the great opportunities that the introduction of machinery must mean for labor in America. Their confidence, however, was not shared by the hand sawyers who, revealing a more jaundiced view of technological improvement as a means to improve economic status, burned Oliver Evans' new steam sawmill at New Orleans in 1813. And intelligent observers like Thomas Cooper reported with misgivings the invasion of the United States by European machines such as the Cartwright loom. Such skepticism and even fear of the deleterious effects of machinery on democratic economic opportunity continued to mid-century. There was no mellow praise for the technologist when, in the thirties, Philadelphia workmen raised a bitter toast to "Labor-saving machinery—Europe's curse and America's dread—Alike the monopolist's idol and the Working Man's scourge—The parent of idleness and consequently crime." A harvesting machine that gave too efficient a demonstration in Michigan in 1847 prompted one pessimistic critic to declaim on "Scientific Improvements a Curse to the Country" and to predict that such so-called improvements would eventually place all economic power in the hands of a few men of capital, relegating labor to the condition of the hapless Irish.

Outstanding among critics of the increasing and uncontrolled production of goods by machinery was Robert Dale Owen whose utopian conscience struggled to weight the correct role of technology within a democratic society. "Mechanical improvements, inevitable even if they were mischievous, and in themselves a rich blessing as sure as they are inevitable—are becoming by some strange perversion of their use, a cruel and deadly curse," Owen warned in 1848. When "the wide and ever-extending west" with its cheap lands was gone, he asked, could the American laborer still hope to remain "a positive quantity"? "If the inventive genius of America, no whit behind that of Europe, brings into being machine after machine . . . is not the laborer, here as in England, thereby liable at last, to be crowded out of the permission to work for his daily bread?" The Owenites, though they prophesied grimly about the machine, did not necessarily wish to eliminate it. Rather, "scientific distribution" as the answer to so-called overproduction became their slogan.

But as faith in industrialism took deeper root in the United States, technology and the ideal of democratic economic opportunity were ever more often equated. English disturbances, it was true, and the sufferings of underpaid or abused machine labor abroad, attracted American sympathy. But there prevailed a strong reaction that "It can't happen here"—America, after all, was a machine-hungry environment. "We do not dread that riot and famine will follow the introduction of such agents into our workshops," proudly reported the chairman of a congressional committee in 1812. He urged instead the great need to import more machinery.

When Charles Knight's new book on *The Results of Machinery*, in part an answer to the anti-machinery riots in England, reached American shores in the early thirties, its generally optimistic points of view attracted wide attention. Knight, with the evidence all around him of a machine civilization older and more advanced than that of the United States, argued that the only basis for the objection to machines lay in simple want of knowledge on that subject. Persuasively, he developed the argument that machinery necessarily benefited mankind in general and the lower economic and social groups in particular. American reviewers picked up specific points of Knight's exposition and editorialized freely upon them. Admitting, as had Knight, the temporary dislocations caused by technological change, one reviewer declared that no improvement in machinery could keep up with the advance in human wants, and that ignorance, vice, and lack of prudence caused the miseries for which machinery was blamed.

The prominent *North American Review* picked up Knight's arguments with relish and enthusiastically recommended the new book "to all croakers,—to all praisers of the past and revilers of the present time. We ask a careful perusal of it, of those venerable grandmothers who see misery and ruin close at hand, because the sound of the spinning-wheel and the loom is no longer heard in all our farm houses." But what about the evidence of widespread British discontent? Blame such discontent, cautioned the writer, not on machinery but on vicious political institutions, unequal laws, grinding taxation. And these, happily, did not really exist in America. Moreover, a more nearly objective view of the whole question was possible in America—industry here was free, labor was guaranteed its wage. Under well-regulated governments machinery always proved to be a blessing; the evils attendant upon its introduction were generally slight and transient, while the benefits "are seen every where, felt every where, and must abide forever." Indeed, concluded the writer, this was a question deserving of more attention in America, but "the general sentiment is decidedly, so far as we have been able to ascertain it, in favor of machinery. A few apostles of the opposite doctrine have arisen here and there; but their converts have not been numerous."

This strong willingness in America to accept the introduction of machinery with such surprising good will, along with the relative absence of labor opposition, perplexed European observers. "Workmen hail with satisfaction all mechanical improvements" reported Joseph Whitworth to the House of Commons in 1854. This labor attitude and "this eager resort to machinery whenever it can be applied," as well as superior education and intelligence, explained, he believed, the remarkable prosperity of the United States.

Democratic economic opportunity, in summary, appears—like social opportunity—to have been advanced rather than retarded by the increasing influence of technology in the United States after 1800. The doubters appeared to be in a minority and remained so until mid-century.

Permeating the concept of democracy in the United States and closely linking its notions of social and economic opportunity with technology, was the current spirit of optimism whch historians have labeled the "idea of progress." So closely were the technologist and his creations associated with progress in America during this period that believers in the relationship seemed often to share a faith more naïve than realistic. Not in terms of religion, nor of philosophy, nor in those of military glory or artistic achievement was the story of human progress to be told. Rather, progress must be identified with the sweeping advances which this age was making in applying science to the satisfaction of human wants. Enthusiastic support for this belief made technology a central element in the optimistic doctrine of progress in America, and by the time of Jackson was shaping for American engineers and inventors a solemn responsibility.

Indeed, by 1830 very many Americans had come to believe that their age was superior to any other, and the prospects of further progress greater, because of remarkable material advancements. Comfort, convenience, and good health were practical evidences of that progress, and constituted, too, a goal for future efforts. Machinery itself became a throbbing symbol of this notion of progress. The writer in the *North American Review* who had found Knight's defense of machinery so deserving had added: "What we claim for machinery is, that it is in modern times by far the most efficient physical cause of human improvement; that it does for civilization, what conquest and human labor formerly did, and accomplishes incalculably more than they accomplished." Edward Everett, then governor of Massachusetts and soon to become president of Harvard, declared in 1837 that "Mind, acting through the useful arts, is the vital principle of modern civilized society. The mechanician, not the magician, is now the master of life."

With so much confidence in the contemporary role of technology, it is not surprising that many Americans saw even greater future challenges for science and invention. Utopia, even, had a technological coloring. Not only did Owen's New Harmony *Gazette* lend ample space to descriptions of new improvements, but the Associationists invited to their ranks men with special scientific and mechanical experience, trusting in the principle of co-operation to control technology in the common interest. It was a modest but calculated surrender to the machine.

Nature's untamed might—steam, water power, electricity—fascinated these nineteenth-century adherents of social progress through technology. What wonders might man not perform by further mastery of wind and tide and the sun's own heat! But it was the glory of steam power, especially, which thrilled Americans and ever greater seemed its promises as each puffing locomotive or steamboat passed by. Not even the compass or the printing press or gunpowder could equal in social significance the impact of steam power, insisted the Reverend James T. Austin in 1839. "It is to bring mankind into a common brotherhood; annihilate space and time in the intercourse of human life; increase the social relations; draw closer the ties of philanthropy and benevolence; multiply common benefits, and the reciprocal interchange of them, and by a power of yet unknown kindness, give to reason and religion an empire which they have but nominally possessed in the conduct of mankind."

Praises and predictions, indeed, went far beyond that of Austin, whether in prose or verse or song. Serious studies tried to define more closely the challenge of exploiting nature's powers. Thomas Ewbank, especially, made his most earnest efforts in such works, coming from his pen after 1840, as *A New Theory of Steam, The Position of Our Species in the Path of Its Destiny*, and *The World a Workshop*. A German-born immigrant inventor, John Adolphus Etzler, imbibed this Yankee enthusiasm and sought a wide audience after 1830 for the descriptions of his carefully blue-printed technological wonderland. Etzler's imagination was vivid, his enthusiasm for the powers of nature unbounded. He addressed "To All Intelligent Men" his detailed prospectus called *The Paradise within the Reach of All Men, without Labor, by Powers of Nature and Machinery*. Etzler outlined his amazing plans to harness tidal action, to subdue the very winds to man's needs, and capture the heat of the sun to power the advanced technological civilization his plans prognosticated. If he inclined to gloss over the *how* of some of the crucial applications by hypothesizing "some simple device," his enthusiasms carried his ideas—and the familiar request for some kind of subsidy—to Congress itself. But such enthusiasms belonged not alone to technically minded planners and dreamers. Even the statesman John C. Calhoun confessed that "the subjugation of electricity to the mechanical necessities of man would mark the lost era in human civilization."

The striking tendency of Americans in these years to measure social progress in material terms brought charges that excessive emphasis on technology was making American democracy dangerously materialistic. What must be the eventual effects of such emphasis on the mind and morals, the religion and aesthetic interests of the average American? Tocqueville, indeed, not only decided that materialism was a dangerous disease of the human mind, but that it was especially to be dreaded in democratic nations. "Democracy encourages a taste for physical gratification: this taste, if it becomes excessive, soon disposes men to believe that all is matter only," he cautioned, and reported that uppermost in every American mind was the determination to satisfy even the least wants of the body and every desire for conveniences.

Some Americans picked up the refrain. "Our philosophy comes from Bacon. It deals only with the wants of man and uses of nature," lamented one of them. Other prominent critics of America's preoccupation with material interests included churchmen like William Ellery Channing and outstanding thinkers like Emerson and Thoreau. The whole question was admirably pointed up by the Reverend Joseph B. Bittinger before an audience at Pennsylvania College in 1860 when he demanded that technologists reexamine their social mission:

> And when our capricious, fictitious and factitious wants demand of the mechanic "Cui bono?" it is not enough for him to say, it will make hats, caps, and shoes, not even if they are waterproof, nor that it will cover umbrellas and make comfortable pillows. The Question to be asked and answered, by man, for man, is, will these improvements tend to make men wiser and better. . . . If man can rise no higher than a mere artisan,

if all his cunning devices and witty inventions shall terminate in time, and be lavished on the body, then will his boasted civilization be a mere Epicureanism . . . his glorious intellect be regarded as no better than "a transcendant mud-wasp, or beaver talent."

Along with such vigorous critiques, of course, appeared also numerous defenses against the charge that technology was breeding gross materialism. One defender protested, for example, that it must be expected that a young nation stoop a little in the path of progress to pick up a few of the material jewels scattered so prodigally at its feet. Some held it unjust to compare the American ideal of improvement for the many with the ascetic or aristocratic concept of the Orient or classical Greece and Rome. When Thomas Carlyle scorned the age as having abandoned the tradition of learning and discovery for the pursuit of gadgetry, American critics hastened to point out that an era of gas lights and lucifer matches could be as enlightened as an era of pine knots and torches. "A chemistry which brews and bakes and cooks for the moderns is quite as philosophical as that which crazed the ancients with a promise of the philosopher's stone," one speaker assured his collegiate audience. "The epoch of steam engines and magnetic telegraphs is no more heretical than the 'devotional ages' when witches were hung, and ghosts exorcised by the priests and then shot with silver bullets."

Nor did the argument rest with a defense of technological materialism. Morally and ethically Americans need anticipate no real danger from science and invention, and should derive great blessings. Mechanical enterprise, after all, provided the freedom and leisure needed to develop man's better nature; the findings of science brought men only a better understanding of, and hence heightened regard for, the handiwork of God. William Barton in 1813 observed in the course of his commentary on the work of the astronomer, David Rittenhouse, that practitioners of this type of "philosophy" must be ranked equal with teachers of religion and morals as disciples of universal truth. Because science enlarged man's views of the operations of Providence in relation to the past and present scenes of the world, clergymen could defend the incessant probing into the secrets of the universe as "but an extensive survey of the empire of God." Let a writer in the *Edinburgh Review* accuse the age as one of technological materialism and an American reader could hasten to reply that this very emphasis on mechanical matters had furnished mankind the freedom and leisure necessary to the true development of the spiritual and immortal part of mankind. Technology, then, a citizen of the nineteenth century might point out, was also a potent weapon against superstition. "Much less, at the present day, when thoughts and words are free, and science is not within the power of intolerance and bigotry, can empty denunciations and misapplied epithets hinder the spirit of progressive investigation and arrest the march of discovery."

Technology's materialistic emphasis was examined not only in the light of its philosophical and spiritual effects, but also for its effect on the intellect of the citizen, especially the machine laborer. Most Americans seemed willing to grant Tocqueville's view that democracies slighted the pursuit of knowledge for its own sake. But would the machine, too, actually dull the imagination and intelligence of the laboring masses? The greatest threat was the specialization of labor which, from the viewpoint of efficiency, minimized skills and boosted production. William Ellery Channing found these attributes scarcely consonant with his own dreams of self-improvement for the laboring class. "The division of labor," cautioned Channing, "tends to dwarf the intellectual powers, by confining the activity of the individual to a narrow range, to a few details, perhaps to the heading of pins, the pointing of nails, to the typing together of broken strings."

But optimistic writers, especially in the technical press, had ready replies to the critics of technology's influence on the laborer's intellect. A reader of the *Working Man's Advocate* in the thirties reminded this Owenite journal that knowledge and consequent enlightenment of public opinion followed in exact proportion the improvements in mechanics and use of artificial power. Meanwhile, the various mechanics' institutes springing up everywhere lured ambitious laborers from the comforts of home or grogshop to venture into areas of knowledge not immediately restricted to the workbench. But when Samuel Colt explained to a British engineering group in 1851 that uneducated laborers made the best workers in his new mass-production arms factory because they had so little to unlearn, there was at least the implication that the machine had introduced a new imponderable in American education.

The life of the mind, the improvement of the spiritual side of man, the expansion of democratic institutions—all these features by 1860 had become a significant part of what many Americans conceived to be the social responsibility of technology in a democratic society. One final point of debate was the relationship between technology and the aesthetic sense.

It was here that Tocqueville made an especially severe criticism of democratic societies and found Americans, especially, wanting in artistic taste or artistic inclination. Quite consistently he linked this weakness with the social mobility, the ambition, the eagerness for immediate results and for physical comforts which he had found so evident. The workman in democracies, said Tocqueville, directed every effort to invent methods which might help him to work not only better but more quickly and cheaply. Sometimes this would mean cutting corners and turning out an inferior product. But if American watches, for example, were vastly inferior to the handful of watches once possessed by aristocrats, at least every American owned a cheap, machine-made watch. Democracy did not necessarily preclude the cultivation of excellent craftsmanship, but few in America were willing to pay the price for such time and trouble, so that on the whole American workmen remained "in a state of accomplished mediocrity." But were not the inferior quality, the abundance, and the cheapness of commodities themselves excellent evidences of a society where privilege and class distinctions were declining? After all, democracies preferred the arts which rendered life easy rather than those which merely adorned it. "They will habitually prefer the useful to the beautiful, and they will require that the beautiful should be useful."

To what degree were Tocqueville's conclusions confirmed by Americans themselves? That expediency might have to take precedence over adornment in a new country had been suggested by Tench Coxe in his admonition to American printers just before the turn of the century to omit the elegancies of fine copperplates and vellum paper because speed of publication was more essential than artistic detail, and because artistic detail "costs more than is agreeable to the people of this country, who desire valuable material for their money." It was only grudgingly that the editors of Fessenden's *Register of Arts* admitted in 1808 that "the cover of the Schuylkill Bridge compelled ornament and some degree of design lest it should disgrace the environs of a great city," however unnecessary such expenditures might be in other locations. And in the 1830s, when the appearance of the American countryside was being altered by the cuts and embankments of the new railroads, the *Boston Mechanics' Magazine* cautioned those who would beautify the new right of way of the Baltimore and Ohio that "Directors and engineers of works of this kind ought to feel that every dollar expended for show and elegance is worse than lost" because it simply cut funds available for other useful projects.

Evidence, however, that the aesthetic impulse was not excluded from technological creation even in the high tide of Jacksonian democracy was to be found in some engineering works both technically and artistically distinguished, and in a modest propaganda for improvement in this relationship. Again, the utilitarian character of the age came under scrutiny. In 1832 the Franklin Institute in Philadelphia reported the deplorable tendency to overornament the manufactures of the age, but noted happily the signs also of a growing competence and style in some of them and advised Americans to produce commodities both simple and neat. The goal of a democratic industrial art, wrote Charles G. Page, a former patent commissioner and now editor of a polytechnic review, in 1854, must be as utilitarian as American science. But this did not exclude an aesthetic purpose. "The form and proportion of every implement or dwelling has a close relation to its purpose," Page instructed, and he praised the American clipper ships as excellent examples of the art in which "the means are directly adapted to the end in view." In American society, Page argued, there was no room for "castellated villas for retired burghers."

In the Crystal Palace exhibition in New York in 1853, Americans could witness personally the proud summing up of a half century of American technological and artistic achievement. Not alone the structure itself but the variety and excellence of the exhibits evoked enthusiasm—the fruits of American engineering and mechanical genius. Graceful form, beautiful outline, and poetic thought, remarked one observer, might now be traced not only in a Madonna or Venus but in "swan-like-boats, those light and airy carriages, and highly finished engines—the epics of mechanics—and the host of tiny-handed operators for sewing, for card making, for spinning, for ornamental weaving, and a multitude of other works, including that wonderful distributor of thought, the printing press." To the Americans, therefore, the New York Crystal Palace exhibition, like the great London international exhibition two years earlier, offered an opportunity to display before the mid-nineteenth-century world a new conception of aesthetic expression. There remained much to be desired in the finish and polish of the samples of machinery and articles of manufacture produced by their own countrymen; but there was no question of the sincerity of purpose and the dedication of those products to the service of Everyman. Their very simplicity and humbleness suggested that.

In a much broader sense, the great exhibitions also summarized the general relationship of technology to the concept of democracy in America, and it seemed appropriate that Walt Whitman should have assisted at the dedication ceremonies in New York in 1853. There was no denying the tremendous technological progress that had been made since 1800—the constant improvement in engineering and the never ending multiplicity of new inventions. As it had begun, so the period ended, with American science and invention still in the service of the people. Popular needs still tended to determine the goals of engineers, inventors, mechanics, and their spokesmen. Perhaps inevitably, the materialistic approach of techonology fitted admirably the tone of life of a bustling people who wanted to get things done quickly, who worshipped abundance, and who believed that every free citizen was entitled to a generous share of the things which brought physical comfort in their world. American technology in the years before the Civil War served those objectives well—a fact of which most Americans seemed fully aware.

Technology and Government

JOHN G. BURKE

One of the most striking characteristics of American society in 1800 was its decentralization. One of the most important results of industrialism was a tendency toward the centralization of power. This was not foreseen by the early spokesmen for modern technology, who assumed that it would be possible to make desired changes, leaving everything else unchanged. In practice things have not worked out that way. Besides the intended results, technology produces unplanned and sometimes undesirable changes. It is often useful to speak of these as primary and secondary effects. Americans were eager for the benefits of technology, but they have become increasingly concerned with the secondary effects. The piecemeal attempts to cope with the secondary effects of technology has led to a vastly more powerful, centralized government. In the following selection, John G. Burke shows how the introduction of the steam engine began this process. He is editor of *The New Technology and Human Values* (Belmont, Calif., 1966) and author of *Origins of the Science of Crystals* (Berkeley, Calif., 1966); and currently he is Dean of Social Sciences at the University of California, Los Angeles.

I

When the United States Food and Drug Administration removes thousands of tins of tuna from supermarket shelves to prevent possible food poisoning, when the Civil Aeronautics Board restricts the speed of certain jets until modifications are completed, or when the Interstate Commerce

John G. Burke, "Bursting Boilers and the Federal Power," *Technology and Culture*, 7 (Winter 1966), 1–23. Copyright © 1966, the Society for the History of Technology. Reprinted without footnotes by permission of the author and the publisher, The University of Chicago Press.

Commission institutes safety checks of interstate motor carriers, the federal government is expressing its power to regulate dangerous processes or products in interstate commerce. Although particular interests may take issue with a regulatory agency about restrictions placed upon certain products or seek to alleviate what they consider to be unjust directives, few citizens would argue that government regulation of this type constitutes a serious invasion of private property rights.

Though federal regulatory agencies may contribute to the general welfare, they are not expressly sanctioned by any provisions of the U.S. Constitution. In fact, their genesis was due to a marked change in the attitude of many early nineteenth-century Americans who insisted that the federal government exercise its power in a positive way in an area that was nonexistent when the Constitution was enacted. At the time, commercial, manufacturing, and business interests were willing to seek the aid of government in such matters as patent rights, land grants, or protective tariffs, but they opposed any action that might smack of governmental interference or control of their internal affairs. The government might act benevolently but never restrictively.

The innovation responsible for the changed attitude toward government regulation was the steam engine. The introduction of steam power was transforming American culture, and while Thoreau despised the belching locomotives that fouled his nest at Walden, the majority of Americans were delighted with the improved modes of transportation and the other benefits accompanying the expanding use of steam. However, while Americans rejoiced over this awesome power that was harnessed in the service of man, tragic events that were apparently concomitant to its use alarmed them—the growing frequency of disastrous boiler explosions, primarily in marine service. At the time, there was not even a governmental agency that could institute a proper investigation of the accidents. Legal definitions of the responsibility or negligence of manufacturers or owners of potentially dangerous equipment were in an embryonic state. The belief existed that the enlightened self-interest of an entrepreneur sufficed to guarantee the public safety. This theory militated against the enactment of any legislation restricting the actions of the manufacturers or users of steam equipment.

Although the Constitution empowered Congress to regulate interstate commerce, there was still some disagreement about the extent of this power even after the decision in *Gibbons* v. *Ogden*, which ruled that the only limitations on this power were those prescribed in the Constitution. In the early years of the republic, Congress passed legislation under the commerce clause designating ports of entry for customs collections, requiring sailing licenses, and specifying procedures for filing cargo manifests. The intent of additional legislation in this area, other than to provide for these normal concomitants of trade, was to promote commerce by building roads, dredging canals, erecting lighthouses, and improving harbors. Congress limited its power under the commerce clause until the toll of death and destruction wrought by bursting steamboat boilers mounted, and some positive regulations concerning the application of steam power seemed necessary. Thomas Jefferson's recommendation that we should have "a wise and frugal Government, which shall restrain men from injuring one another, shall leave them otherwise free to regulate their own pursuits of industry and improvement" took on a new meaning.

Although several historians have noted the steamboat explosions and the resulting federal regulations, the wider significance of the explosions as an important factor in altering the premises concerning the role of government vis à vis private enterprise has been slighted. Further, there has been no analysis of the role of the informed public in this matter. The scientific and technically knowledgeable members of society were—in the absence of a vested interest—from the outset firmly committed to the necessity of federal intervention and regulation. They conducted investigations of the accidents; they proposed detailed legislation which they believed would prevent the disasters. For more than a generation, however, successive Congresses hesitated to take forceful action, weighing the admitted danger to the public safety against the unwanted alternative, the regulation of private enterprise.

The regulatory power of the federal government, then, was not expanded in any authoritarian manner. Rather, it evolved in response to novel conditions emanating from the new machine age, which was clearly seen by that community whose educations or careers encompassed the new technology. In eventually reacting to this danger, Congress passed the first positive regulatory legislation and created the first agency empowered to supervise and direct the internal affairs of a sector of private enterprise in detail. Further, certain congressmen used this precedent later in efforts to protect the public in other areas, notably in proposing legislation that in time created the Interstate Commerce Commission. Marine boiler explosions, then, provoked a crisis in the safe application of steam power, which led to a marked change in American political attitudes. The change, however, was not abrupt but evolved between 1816 and 1852.

11

Throughout most of the eighteenth century, steam engines worked on the atmospheric principle. Steam was piped to the engine cylinder at atmospheric pressure, and a jet of cold water introduced into the cylinder at the top of the stroke created a partial vacuum in the cylinder. The atmospheric pressure on the exterior of the piston caused the power stroke. The central problem in boiler construction, then, was to prevent leakage. Consequently, most eighteenth-century boilers were little more than large wood, copper, or cast-iron containers placed over a hearth and encased with firebrick. In the late eighteenth century, Watt's utilization of the expansive force of steam compelled more careful boiler design. Using a separate condenser in conjunction with steam pressure, Watt operated his engines at about 7 p.s.i. above that of the atmosphere. Riveted wrought-iron boilers were introduced, and safety valves were employed to discharge steam if the boiler pressure exceeded the designed working pressure.

Oliver Evans in the United States and Richard Trevithick in England introduced the relatively high-pressure non-condensing steam engine almost simultaneously at the turn of the nineteenth century. This development led to the vast extension in the use of steam power. The high-pressure engines competed in efficiency with the low-pressure type, while their compactness made them more suitable for land and water vehicular transport. But, simultaneously, the scope of the problem faced even by Watt was increased, that is, the construction of boilers that would safely contain the dangerous expansive force of steam. Evans thoroughly respected the potential destructive force of steam. He relied chiefly on safety valves with ample relieving capacity but encouraged sound boiler design by publishing the first formula for computing the thickness of wrought iron to be used in boilers of various diameters carrying different working pressures.

Despite Evans' prudence, hindsight makes it clear that the rash of boiler explosions from 1816 onward was almost inevitable. Evans' design rules were not heeded. Shell thickness and diameter depended upon available material, which was often of inferior quality. In fabrication, no provision was made for the weakening of the shell occasioned by the rivet holes.

The danger inherent in the employment of wrought-iron shells with cast-iron heads affixed because of the different coefficients of expansion was not recognized, and the design of internal stays was often inadequate. The openings in the safety valves were not properly proportioned to give sufficient relieving capacity. Gauge cocks and floats intended to ensure adequate water levels were inaccurate and subject to malfunction by fouling with sediment or rust.

In addition, there were also problems connected with boiler operation and maintenance. The rolling and pitching of steamboats caused alternate expansion and contraction of the internal flues as they were covered and uncovered by the water, a condition that contributed to their weakening. The boiler feedwater for steamboats was pumped directly from the surroundings without treatment or filtration, which accelerated corrosion of the shell and fittings. The sediment was frequently allowed to accumulate, thus requiring a hotter fire to develop the required steam pressure, which led, in turn, to a rapid weakening of the shell. Feed pumps were shut down at intermediate stops without damping the fires, which aggravated the danger of low water and excessive steam pressure. With the rapid increase in the number of steam engines, there was a concomitant shortage of competent engineers who understood the necessary safety precautions. Sometimes masters employed mere stokers who had only a rudimentary of the operation of steam equipment. Increased competition also led to attempts to gain prestige by arriving first at the destination. The usual practice during a race was to overload or tie down the safety valve, so that excessive steam pressure would not be relieved.

III

The first major boiler disasters occurred on steamboats, and, in fact, the majority of explosions throughout the first half of the nineteenth century took place on board ship. By mid-1817, four explosions had taken five lives in the eastern waters, and twenty-five people had been killed in three accidents on the Ohio and Mississippi rivers. The city council of Philadelphia appears to have been the first legislative body in the United States to take cognizance of the disasters and attempt an investigation. A joint committee was appointed to determine the causes of the accidents and recommend measures that would prevent similar occurrences on steamboats serving Philadelphia. The question was referrred to a group of practical engineers who recommended that all boilers should be subjected to an initial hydraulic proof test at twice the intended working pressure and additional monthly proof tests to be conducted by a competent inspector. Also, appreciating the fact that marine engineers were known to overload the safety valve levers, they advocated placing the valve in a locked box. The report of the joint committee incorporated these recommendations, but it stated that the subject of regulation was outside the competence of municipalities. Any municipal enactment would be inadequate for complete regulation. The matter was referred, therefore, to the state legislature, and there it rested.

Similar studies were being undertaken abroad. In England, a fatal explosion aboard a steamboat near Norwich prompted Parliament to constitute a Select Committee in May 1817 to investigate the conditions surrounding the design, construction, and operation of steam boilers. In its report, the committee noted its aversion to the enactment of any legislation but stated that where the public safety might be endangered by ignorance, avarice, or inattention, it was the duty of Parliament to interpose. Precedents for legislation included laws covering the construction of party walls in buildings, the qualification of physicians, and the regulation of stage coaches. The committee recommended that passenger-carrying steam vessels should be registered, that boiler construction and testing should be supervised, and that two safety valves should be employed with severe penalties for tampering with the weights.

No legislation followed this report, nor were any laws enacted after subsequent reports on the same subject in 1831, 1839, and 1843. The attitude of the British steamboat owners and boiler manufacturers was summarized in a statement that the prominent manufacturer, Sir John Rennie, made to the Select Committee in 1843. There should be, he said, no impediments in the application of steam power. Coroners' juries made such complete investigations of boiler explosions that no respectable manufacturer would risk his reputation in constructing a defective boiler. Constant examination of boilers, he argued, would cause serious inconvenience and would give no guarantee that the public safety would be assured. Admittedly, it would be desirable for steam equipment to be perfect, but with so many varied boiler and engine designs, it would be next to impossible to agree on methods of examination. Besides, he concluded, there were really few accidents.

In this latter remark, Sir John was partially correct. In England, from 1817 to 1839, only 77 deaths resulted from twenty-three explosions. This record was relatively unblemished compared to the slaughter in the United States, where in 1838 alone, 496 lives had been lost as a result of fourteen explosions. The continued use of low-pressure engines by the British; the fact that by 1836 the total number of U.S. steamboats—approximately 750 —was greater than the total afloat in all of Europe; and the fact that the average tonnage of U.S. steamboats was twice that of British vessels, implying the use of larger engines and boilers and more numerous passengers, accounted for the large difference in the casualty figures.

In France, the reaction to the boiler hazard was entirely different than in Great Britain and the United States. Acting under the authority of Napoleonic legislation, the government issued a Royal Ordinance on October 29, 1823 relative to stationary and marine steam engines and boilers. A committee of engineers of mines and civil engineers prepared the regulations, but the scientific talent of such men as Arago, Dulong, and Biot was enlisted to prepare accurate steam tables. By 1830, amendments resulted in the establishment of a comprehensive boiler code. It incorporated stress values for iron and copper and design formulas for these materials. It required the use of hemispherical heads on all boilers operating above 7 p.s.i. and the employment of two safety valves, one of which was enclosed in a locked grating. Boiler shells had to be fabricated with fusible metal plates made of a lead-tin-bismuth alloy and covered with a cast-iron grating to prevent swelling when close to the fusing point. Boilers had to be tested initially at three times the designed working pressure and yearly thereafter. The French engineers of mines and government civil engineers were given detailed instructions on the conduct the tests and were empowered to remove any apparently defective boiler from service. The proprietors of steamboats or factories employing boilers were liable to criminal prosecution for evasion of the regulations, and the entire hierarchy of French officialdom was enjoined to report any infractions.

Proper statistics proving that this code had a salutary effect in the prevention of boiler explosions are not available. It is certain that some explosions occurred despite the tight regulations. Arago, writing in 1830, reported that a fatal explosion on the "Rhone" resulted from the tampering with a safety valve and pointed out that fusion of the fusible metal plates could be prevented by directing a stream of water on them. Undoubtedly, in some instances the laws were evaded, but Thomas P. Haldeman, an experienced Cincinnati steamboat captain said in 1848 that the code had been effective. He wrote: "Since those laws were enforced we have scarcely heard of an explosion in that country. ... What a misfortune our government did not follow the example of France twenty years ago." Significantly, both Belgium and Holland promulgated boiler laws that were in all essentials duplicates of the French regulations.

IV

From 1818 to 1824 in the United States, the casualty figures in boiler disasters rose, about forty-seven lives being lost in fifteen explosions. In May 1824 the "Aetna," built in 1816 to Evans' specifications, burst one of her three wrought-iron boilers in New York harbor, killing about thirteen persons and causing many injuries. Some experts attributed the accident to a stoppage of feedwater due to incrustations in the inlet pipes, while others believed that the rupture in the shell had started from an old fracture in a riveted joint. The accident had two consequences. Because the majority of steamboats plying New York waters operated at relatively low pressures with copper boilers, the public became convinced that wrought-iron boilers were unsafe. This prejudice forced New York boat builders who were gradually recognizing the superiority of wrought iron to revert to the use of copper even in high-pressure boilers. Some owners recognized the danger of this step, but the outcry was too insistent. One is reported to have said: "We have concluded therefore to give them [the public] a copper boiler, the strongest of its class, and have made up our minds that they have a perfect right to be scalded by copper boilers if they insist upon it." His forecast was correct, for within the next decade, the explosion of copper boilers employing moderate steam pressures became common in eastern waters.

The second consequence of the "Aetna" disaster was that it caught the attention of Congress. A resolution was introduced in the House of Representatives in May 1824 calling for an inquiry into the expediency of enacting legislation barring the issuance of a certificate of navigation to any boat operating at high steam pressures. Although a bill was reported out of committee, it was not passed due to lack of time for mature consideration.

In the same year, the Franklin Institute was founded in Philadelphia for the study and promotion of the mechanical arts and applied science. The institute soon issued its *Journal*, and, from the start, much space was devoted to the subject of boiler explosions. The necessity of regulatory legislation dealing with the construction and operation of boilers was discussed, but there was a diversity of opinion as to what should be done. Within a few years, it became apparent that only a complete and careful investigation of the causes of explosions would give sufficient knowledge for suggesting satisfactory regulatory legislation. In June 1830, therefore, the Institute empowered a committee of its members to conduct such an investigation and later authorized it to perform any necessary experiments.

The statement of the purpose of the committee reflects clearly the nature of the problem created by the frequent explosions. The public, it said, would continue to use steamboats, but if there were no regulations, the needless waste of property and life would continue. The committee believed that these were avoidable consequences; the accidents resulted from defective boilers, improper design, or carelessness. The causes, the committee thought, could be removed by salutary regulations, and it affirmed: "That there must be a power in the community lodged somewhere, to protect the people at large against any evil of serious and frequent recurrence, is self-evident. But that such power is to be used with extreme caution, and only when the evil is great, and the remedy certain of success, seems to be equally indisputable."

Here is a statement by a responsible group of technically oriented citizens that public safety should not be endangered by private negligence. It demonstrates the recognition that private enterprise was considered sacrosanct, but it calls for a reassessment of societal values in the light of events. It proposes restrictions while still professing unwillngness to fetter private industry. It illustrates a change in attitude that was taking place with respect to the role of government in the affairs of industry, a change that was necessitated by technological innovation. The committee noted that boiler regulation proposals had been before Congress twice without any final action. Congressional committees, it said, appeared unwilling to institute inquiries and elicit evidence from practical men, and therefore they could hardly determine facts based upon twenty years of experience with the use of steam in boats. Since Congress was apparently avoiding action, the committee asserted, it was of paramount importance that a competent body whose motives were above suspicion should shoulder the burden. Thus, the Franklin Institute committee began a six-year investigation of boiler explosions.

From 1825 to 1830, there had been forty-two explosions killing about 273 persons, and in 1830 a particularly serious one aboard the "Helen McGregor" near Memphis which killed 50 or 60 persons, again disturbed Congress. The House requested the Secretary of the Treasury, Samuel D. Ingham of Pennsylvania, to investigate the boiler accidents and submit a report. Ingham had served in Congress from 1813 to 1818, and again from 1822 to 1829. He was a successful manufacturer who owned several paper mills; he was acquainted with the activities of the Franklin Institute and had written to the *Journal* about steam boiler problems. Ingham was thus in a unique position to aid the Franklin Institute committee which had begun its inquiries. Before his resignation from Jackson's cabinet over the Peggy O'Neil Eaton affair, Ingham committed government funds to the Institute to defray the cost of apparatus necessary for the experiments. This was the first research grant of a technological nature made by the federal government.

Ingham attempted to make his own investigation while still secretary of the treasury. His interim report to the House in 1831 revealed that two investigators, one on the Atlantic seaboard and the other in the Mississippi basin, had been employed to gather information on the boiler explosions. They complained that owners and masters of boats seemed unwilling to aid the inquiry. They were told repeatedly that the problem was purely individual, a matter beyond the government's right to interfere. In the following year, the new secretary, Louis B. McLane, circulated a questionnaire among the collectors of customs, who furnished information and solicited opinions about the explosions. Their answers formed the basis of McLane's report to Congress. They mentioned the many causes of boiler explosions. One letter noted that steamboat trips from New Orleans to Louisville had been shortened from twenty-five to twelve days since 1818 without increasing the strength of the boilers. A frequent remark was that the engineers in charge of the boilers were ignorant, careless, and usually drunk.

This report prompted a bill proposed in the House in May 1832. It provided for the appointment of inspectors at convenient locations to test the strength of the boilers every three months at three times their working pressure, and the issuance of a license to navigate was made contingent upon this inspection. To avoid possible objections on the score of expense, inspection costs were to be borne by the government. To prevent explosions caused by low water supply, the bill provided that masters and engineers be required under threat of heavy penalties to supply water to the boilers while the boat was not in motion.

The half-hearted tone of the House committee's report on the bill hardly promised positive legislative action. The Constitution gave Congress the power to regulate commerce, the report noted, but the right of Congress to prescribe the mode, manner, or form of construction of the vehicles of conveyance could not be perceived. Whether boats should be propelled by wind, paddles, or steam, and if by steam, whether by low or high pressure, were questions that were not the business of Congress. No legislation was competent to remove the causes of boiler explosions, so that steam and its application must be left to the control of intellect and practical science. The intelligent conduct of those engaged in its use would be the best safeguard against the dangers incident to negligence. Besides, the report concluded, the destruction was much less than had been thought; the whole number of explosions in the United States was only fifty-two, with total casualties of 256 killed and 104 injured. Supporters of the bill could not undo the damage of the watered-down committee report, however. The bill died, and the disasters continued.

In his State of the Union message in December 1833, President Jackson noted that the distressing accidents on steamboats were increasing. He suggested that the disasters often resulted from criminal negligence by masters of the boats and operators of the engines. He urged Congress to pass precautionary and penal legislation to reduce the accidents. A few days later, Senator Daniel Webster proposed that the Committee on Naval Affairs study the problem. He suggested that all boilers be tested at three times their working pressure and that any steamboat found racing be forfeited to the government. Thomas Hart Benton followed Webster, stating that the matter properly was the concern of the Judiciary Committee. The private waters of states were involved, Benton said; interference with their sovereignty might result. In passing, Benton remarked that the masters and owners of steamboats were, with few exceptions, men of the highest integrity. Further, Benton said, *he* had never met with any accident on a steamboat despite the fact that he traveled widely; upon boarding he was always careful to inquire whether the machinery was in good order. Webster still carried the day, since the matter went to the Committee on Naval Affairs; however, Benton's attitude prevailed in the session, for the reported Senate bill failed to pass.

V

A program of experiments carried out by the Franklin Institute from 1831 to 1836 was based largely upon the reports of circumstances surrounding previous boiler explosions, the contemporary design and construction of boilers and their accessories, and methods of ensuring an adequate water supply. The work was done by a committee of volunteers led by Alexander Dallas Bache, later superintendent of the U.S. Coast Survey, who, at the time, was a young professor of natural philosophy at the University of Pennsylvania. A small boiler, one foot in diameter and about three feet long, with heavy glass viewing ports at each end, was used in most of the experiments. In others, the zeal of the workers led them to cause larger boilers to burst at a quarry on the outskirts of Philadelphia.

The group's findings overturned a current myth, proving conclusively that water did not decompose into hydrogen and oxygen inside the boiler, with the former gas exploding at some high temperature. The experimenters demonstrated that an explosion could occur without a sudden increase of pressure. Another widely held theory they disproved was that when water was injected into a boiler filled with hot and unsaturated steam, it flashed into an extremely high-pressure vapor, which caused the boiler to rupture. The group proved that the reverse was true: the larger the quantity of water thus introduced, the greater the decrease in the steam pressure.

The Franklin Institute workers also produced some positive findings. They determined that the gauge cocks, commonly used to ascertain the level of water inside the boilers, did not in fact show the true level, and that a glass tube gauge was much more reliable, if kept free from sediment. They found the fusing points of alloys of lead, tin, and bismuth, and recommended that fusible plates be employed with caution, because the more fluid portion of an alloy might be forced out prior to the designated fusion temperature, thus leaving the remainder with a higher temperature of fusion. They investigated the effect of the surface condition of the shell on the temperature and time of vaporization, and they determined that properly weighted safety valves opened at calculated pressures within a small margin of error. The results of their experiments on the relationship of the pressure and temperature of steam showed close correspondence with those of the French, although, at this time, values of the specific heat of steam were erroneous due to the inability to differentiate between constant volume and constant temperature conditions.

Simultaneously, another committee, also headed by Bache, investigated the strength of boiler materials. In these experiments, a sophisticated tensile testing machine was constructed, and corrections were made for friction and stresses produced during the tests. The investigators tested numerous specimens of rolled copper and wrought iron, not only at ambient temperatures but up to 1,300° F. They showed conclusively that there were substantial differences in the quality of domestic wrought irons by the differences in yield and tensile strengths. Of major importance was their finding that there was a rapid decrease in the ultimate strength of copper and wrought iron with increasing temperature. Further, they determined that the strength of iron parallel to the direction of rolling was about 6 percent greater than in the direction at right angles to it. They proved that the laminated structure in "piled" iron, forged from separate pieces, yielded much lower tensile values than plate produced from single blooms. Their tests also showed that special precautions should be taken in the design of riveted joints.

Taken as a whole, the Franklin Institute reports demonstrate remarkable experimental technique as well as a thorough methodological approach. They exposed errors and myths in popular theories on the nature of steam and the causes of explosions. They laid down sound guidelines on the choice of materials, on the design and construction of boilers, and on the design and arrangement of appurtenances added for their operation and safety. Further, the reports included sufficient information to emphasize the necessity for good maintenance procedures and frequent proof tests, pointing out that the strength of boilers diminished as the length of service increased.

VI

The Franklin Institute report on steam boiler explosions was presented to the House through the secretary of the treasury in March 1836, and the report on boiler materials was available in 1837. The Franklin Institute committee also made detailed recommendations on provisions that any regulatory legislation should incorporate. It proposed that inspectors be appointed to test all boilers hydraulically every six months; it prohibited the licensing of ships using boilers whose design had proved to be unsafe; and it recommended penalties in cases of explosions resulting from improper maintenance, from the incompetence or negligence of the master or engineer, or from racing. It placed responsibility for injury to life or property on owners who neglected to have the required inspections made, and it recommended that engineers meet certain standards of experience, knowledge, and character. The committee had no doubt of the right of Congress to legislate on these matters.

Congress did not act immediately. In December 1836 the House appointed a committee to investigate the explosions, but there was no action until after President Van Buren urged the passage of legislation in December 1837. That year witnessed a succession of marine disasters. Not all were attributable to boiler explosions, although the loss of 140 persons in a new ship, the "Pulaski," out of Charleston, was widely publicized. The Senate responded quickly to Van Buren's appeal, passing a measure on January 24, 1838. The House moved less rapidly. An explosion aboard the "Moselle" at Cincinnati in April 1838, which killed 151 persons, caused several Congressmen to request suspension of the rules so that the bill could be brought to the floor, but in the face of more pressing business the motion was defeated. The legislation was almost caught in the logjam in the House at the end of the session, but on June 16 the bill was brought to the floor. Debate centered principally upon whether the interstate commerce clause in the Constitution empowered Congress to pass such legislation. Its proponents argued affirmatively, and the bill was finally approved and became law on July 7, 1838.

The law incorporated several sections relating to the prevention of collisions, the control of fires, the inspection of hulls, and the carrying of lifeboats. It provided for the immediate appointment by each federal judge of a competent boiler inspector having no financial interest in their manufacture. The inspector was to examine every steamboat boiler in his area semiannually, ascertain its age and soundness, and certify it with a recommended working pressure. For this service the owner paid the inspector $5.00—his sole remuneration—and a license to navigate was contingent upon the receipt of this certificate. The law specified no inspection criteria. It enjoined the owners to employ a sufficient number of competent and experienced engineers, holding the owners responsible for loss of life or property damage in the event of a boiler explosion for their failure to do so. Further, any steamboat employee whose negligence resulted in the loss of life was to be considered guilty of manslaughter, and upon conviction could be sentenced to not more than ten years imprisonment. Finally, it provided that in suits against owners for damage to persons or property, the fact of the bursting of the boilers should be considered prima facie evidence of negligence until the defendant proved otherwise.

This law raises several questions, because the elimination of inspection criteria and the qualification of engineers rendered the measure ineffectual. Why was this done? Did Congress show restraint because it had insufficient information? Did it yield to the pressure of steamboat interests who feared government interference? Such questions cannot be definitely answered, but there are clues for some tentative conclusions.

The bill, as originally introduced, was similar to the Franklin Institute proposals, so that the Senate committee to which it was referred possessed the most recent informed conclusions as to the causes of boiler explosions and the means of their prevention. The President's plea to frame legislation in the face of the mounting fatalities undoubtedly persuaded the Democratic majority to act. They were unmoved by a memorial from steamboat interests urging the defeat of the bill. But the majority was not as yet prepared to pass such detailed regulations as had originally been proposed. In response to a question as to why the provision for the qualification of engineers had been eliminated, the Senate committee chairman stated that the committee had considered this requirement desirable but foresaw too much difficulty in putting it into effect. Further, the Senate rejected an amendment to levy heavy penalties for racing, as proposed by the Whig, Oliver Smith of Indiana. The Whigs appear to have seen the situation as one in which the federal government should use its powers and interpose firmly. Henry Clay, R. H. Bayard of Delaware, and Samuel Prentiss of Vermont supported Smith's amendment, and John Davis of Massachusetts declared that he would support the strongest measures to make the bill effective. Those who had urged rapid action of the bill in the House were William B. Calhoun and Caleb Cushman of Massachusetts and Elisha Whittlesey of Ohio, all Whigs. But at this time the majority hewed to the doctrine that enlightened self-interest should motivate owners to provide safe operation. The final clause, specifying that the bursting of boilers should be taken as prima facie evidence of negligence until proved otherwise, stressed this idea.

The disappointment of the informed public concerning the law was voiced immediately in letters solicited by the secretary of the treasury, contained in a report that he submitted to Congress in December 1838. There were predictions that the system of appointment and inspection would encourage corruption and graft. There were complaints about the omission of inspection criteria and a provision for the licensing of engineers. One correspondent pointed out that it was impossible legally to determine the experience and skill of an engineer, so that the section of the law that provided penalties for owners who failed to employ experienced and skilful engineers was worthless. One critic who believed that business interests had undue influence upon the government wrote: "We are mostly ruled by corporations and joint-stock companies. . . . If half the citizens of this country should get blown up, and it should be likely to affect injuriously the trade and commerce of the other half by bringing to justice the guilty, no elective officer would risk his popularity by executing the law."

But there also was a pained reaction from the owners of steamboats. A memorial in January 1841 from steamboat interests on the Atlantic seaboard stressed that appropriate remedies for the disasters had not been afforded by the 1838 law as evidenced by the casualty figures for 1839 and 1840. They provided statistics to prove that in *their* geographical area the loss of life per number of lives exposed had decreased by a factor of sixteen from 1828 to 1838, indicating that the troubles centered chiefly in the western waters. But at the same time the memorial emphasized that the 1838 law acted as a deterrent for prudent men to continue in the steamboat business, objecting particularly to the clause that construed a fatal disaster as prime facie evidence of negligence. They argued that if Congress considered steam navigation too hazardous for the public safety, it would be more just and honorable to prohibit it entirely.

However, it not only was the Congress that was reconsidering the concepts of negligence and responsibility in boiler explosions. The common law also searched for precedents to meet the new conditions, to establish guidelines by which to judge legal actions resulting from technological innovation. A key decision, made in Pennsylvania in 1845, involved a boiler explosion at the defendant's flour mill that killed the plaintiff's horse. The defense pleaded that any negligence was on the part of the boiler manufacturer. The court, however, ruled otherwise, stating that the owner of a public trade or business which required the use of a steam engine was responsible for any injury resulting from its deficiency. This case was used as a precedent in future lawsuits involving boiler explosions.

VII

Experience proved that the 1838 law was not preventing explosions or loss of life. In the period 1841–48, there were some seventy marine explosions that killed about 625 persons. In December 1848 the commissioner of patents, to whom Congress now turned for data, estimated that in the period 1816–48 a total of 233 steamboat explosions had occurred in which 2,563 persons had been killed and 2,097 injured, with property losses in excess of $3 million.

In addition to the former complaints about the lack of proof tests and licenses for engineers, the commissioner's report included testimony that the inspection methods were a mockery. Unqualified inspectors were being appointed by district judges through the agency of highly placed friends. The inspectors regarded the position as a lifetime office. Few even looked at the boilers but merely collected their fees. The inspector at New York City complained that his strict inspection caused many boats to go elsewhere for inspections. He cited the case of the "Niagara," plying between New York City and Albany, whose master declined to take out a certificate from his office because it recommended a working pressure of only 25 p.s.i. on the boiler. A few months later the boiler of the "Niagara," which had been certified in northern New York, exploded while carrying a pressure of 44 p.s.i. and killed two persons.

Only eighteen prosecutions had been made in ten years under the manslaughter section of the 1838 law. In these cases there had been nine convictions, but the penalties had, for the most part, been fines which were remitted. It was difficult to assemble witnesses for a trial, and juries could not be persuaded to convict a man for manslaughter for an act of negligence, to which it seemed impossible to attach this degree of guilt. Also, the commissioner's report pointed out that damages were given in cases of bodily injury but that none was awarded for loss of life in negligence suits. It appeared that exemplary damages might be effective in curbing rashness and negligence.

The toll of life in 1850 was 277 dead from explosions, and in 1851 it rose to 407. By this time Great Britain had joined France in regulatory action, which the Congress noted. As a consequence of legislation passed in 1846 and 1851, a rejuvenated Board of Trade was authorized to inspect steamboats semiannually, to issue or deny certificates of adequacy, and to investigate and report on accidents. The time had come for the Congress to take forceful action, and in 1852 it did.

John Davis, Whig senator from Massachusetts, who had favored stricter legislation in 1838, was the driving force behind the 1852 law. In prefacing his remarks on the general provisions of the bill, he said: "A very extensive correspondence has been carried on with all parts of the country . . . there have been laid before the committee a great multitude of memorials, doings of chambers of commerce, of boards of trade, of conventions, of bodies of engineers; and to a considerable extent of all persons interested, in one form or another, in steamers . . . in one thing . . . they are all . . . agreed—that is, that the present system is erroneous and needs correction."

Thus again, the informed public submitted recommendations on the detailed content of the measure. An outstanding proponent who helped shape the bill was Alfred Guthrie, a practical engineer from Illinois. With personal funds, Guthrie had inspected some two hundred steamboats in the Mississippi valley to ascertain the causes of boiler explosions. Early in the session, Senator Shields of Illinois succeeded in having Guthrie's report printed, distributed, and included in the Senate documents. Guthrie's recommendations were substantially those made by the Franklin Institute in 1836. His reward was the post as first supervisor of the regulatory agency which the law created.

After the bill reached the Senate floor, dozens of amendments were proposed, meticulously scrutinized, and disposed of. The measure had been, remarked one senator, "examined and elaborated . . . more patiently, thoroughly, and faithfully than any other bill before in the Senate of the United States." As a result, in place of the 1838 law which embodied thirteen sections and covered barely three pages, there was passed such stringent and restrictive legislation that forty-three sections and fourteen pages were necessary.

The maximum allowable working pressure for any boiler was set at 110 p.s.i., and every boiler had to be tested yearly at one and one-half times its working pressure. Boilers had to be fabricated from suitable quality iron plates, on which the manufacturer's name was stamped. At least two ample safety valves—one in a locked grating—were required, as well as fusible plates. There were provisions relating to adequate supply of boiler feedwater and outlawing designs that might prove dangerous. Inspectors were authorized to order repairs at any time. All engineers had to be licensed by inspectors, and the inspectors themselves issued certificates only under oath. There were stiff monetary penalties for any infractions. The penalty for loading a safety valve excessively was a two hundred dollar fine and eighteen months imprisonment. The fine for manufacturing or using a boiler of unstamped material was five hundred dollars. Fraudulent stamping carried a penalty of five hundred dollars and two years imprisonment. Inspectors falsifying certificates were subject to a five hundred dollar fine and six months imprisonment, and the law expressly prohibited their accepting bribes.

A new feature of the law, which was most indicative of the future, was the establishment of boards of inspectors empowered to investigate infractions or accidents, with the right to summon witnesses, to compel their attendance, and to examine them under oath. Above the local inspectors were nine supervisors appointed by the President. Their duties included the compilation of evidence for the prosecution of those failing to comply with the regulations and the preparation of reports to the secretary of the treasury on the effectiveness of the regulations. Nor did these detailed regulations serve to lift the burden of presumptive negligence from the shoulders of owners in cases of explosion. The explosion of boilers was not made prime facie evidence as in the 1838 law, but owners still bore a legal responsibility. This was made clear in several court decisions which held that proof of strict compliance with the 1852 law was not a sufficient defense to the allegations of loss by an explosion caused by negligence.

The final Senate debate and the vote on this bill shows how, in thirty years, the public attitude and, in turn, the attitudes of its elected representatives had changed toward the problem of unrestricted private enterprise, mainly as a result of the boiler explosions. The opponents of the bill still argued that the self-interest of the steamboat companies was the best insurance of the safety of the traveling public. But their major argument against passage was the threat to private property rights which they considered the measure entailed. Senator Robert F. Stockton of New Jersey was most emphatic:

> It is this—how far the Federal Government . . . shall be permitted to interfere with the rights of personal property—or the private business of any citizen . . . under the influence of recent calamities, too much

sensibility is displayed on this subject . . . I hold it to be my imperative duty not to permit my feelings of humanity and kindness to interfere with the protection which I am bound, as a Senator of the United States, to throw around the liberty of the citizen, and the investment of his property, or the management of his own business . . . what will be left of human liberty if we progress on this course much further? What will be, by and by, the difference between citizens of this far-famed Republic and the serfs of Russia? Can a man's property be said to be his own, when you take it out of his own control and put it into the hands of another, though he may be a Federal officer?

This expression of a belief that Congress should in no circumstances interfere with private enterprise was now supported by only a small minority. One proponent of the bill replied: "I consider that the only question involved in the bill is this: Whether we shall permit a legalized, unquestioned, and peculiar class in the community to go on committing murder at will, or whether we shall make such enactments as will compel them to pay some attention to the value of life." It was, then, a question of the sanctity of private property rights as against the duty of government to act in the public weal. On this question the Senate voted overwhelmingly that the latter course should prevail.

Though not completely successful, the act of 1852 had the desired corrective effects. During the next eight years prior to the outbreak of the Civil War, the loss of life on steamboats from all types of accidents dropped to 65 percent of the total in the corresponding period preceding its passage. A decade after the law became effective, John C. Merriam, editor and proprietor of the *American Engineer*, wrote: "Since the passage of this law steamboat explosions on the Atlantic have become almost unknown, and have greatly decreased in the west. With competent inspectors, this law is invaluable, and we hope to hail tht day when a similar act is passed in every legislature, touching locomotive and stationary boilers.

There was, of course, hostility and opposition to the law immediately after its passage, particularly among the owners and masters of steamboats. It checked the steady rise in the construction of new boats, which had been characteristic of the earlier years. The effect, however, was chastening rather than emasculating. Associations for the prevention of steam boiler explosions were formed; later, insurance companies were organized to insure steam equipment that was manufactured and operated with the utmost regard for safety. In time, through the agency of the American

Society of Mechanical Engineers, uniform boiler codes were promulgated and adopted by states and municipalities.

Thus, the reaction of the informed public, expressed by Congress, to boiler explosions caused the initiation of positive regulation of a sector of private enterprise through a governmental agency. The legislation reflected a definite change of attitude concerning the responsibility of the government to interfere in those affairs of private enterprise where the welfare and safety of the general public was concerned. The implications of this change for the future can be seen by reference to the Windom Committee report of 1874, which was the first exhaustive study of the conditions in the railroad industry that led ultimately to the passage of legislation creating the Interstate Commerce Commission. One section of this report was entitled: "The Constitutional Power of Congress to Regulate Commerce among the Several States." The committee cited the judicial interpretation of the Constitution in *Gibbons* v. *Ogden*, that it was the prerogative of Congress solely to regulate interstate commerce, and also referred to the decision of Chief Justice Taney in *Genesee Chief* v. *Fitzhugh*, wherein it was held that this power was as extensive upon land as upon water. The report pointed out that no decision of the Supreme Court had ever countenanced the view that the power of Congress was purely negative, that it could be constitutionally exercised only by disburdening commerce, by preventing duties and imposts on the trade between the states. It fact, the report argued, Congress had already asserted its power positively. Referring to the acts of 1838 and 1852, it stated that "Congress has passed statutes defining how steamboats shall be constructed and equipped." Thus, the legislation that was provoked by bursting boilers was used as a precedent to justify regulatory legislation in another area where the public interest was threatened.

Bursting steamboat boilers, then, should be viewed not merely as unfortunate and perhaps inevitable consequences of the early age of steam, as occurrences which plagued nineteenth-century engineers and which finally, to a large degree, they were successful in preventing. They should be seen also as creating a dilemma as to how far the lives and property of the general public might be endangered by unrestricted private enterprise. The solution was an important step toward the inauguration of the regulatory and investigative agencies in the federal government.

Alienation and Technology

LEO MARX

The enthusiasm that many Americans once felt for technology has cooled. Today opposition is widespread, and technology is blamed for modern alienation. Literary figures, as Leo Marx points out, were in the vanguard of this change in national mood. Is this growing disenchantment a result of ignorance? Do the critics see only the harmful effects of technology and not its benefits? Or, as Marx suggests, do sensitive spirits react against technology's apparent lack of moral purpose? Whatever the cause, the alienation of a growing portion of the population is a serious problem. Leo Marx is the author of *The Machine in the Garden: Technology and the Pastoral Ideal in America* (New York, 1964) and a professor of English at Amherst College.

> . . . the artist must employ the symbols in use in his day and nation to convey his enlarged sense to his fellow-men.
>
> RALPH WALDO EMERSON

I

The response of American writers to industrialism has been a typical and, in many respects, a distinguishing feature of our culture. The Industrial Revolution, of course, was international, but certain aspects of the process were intensified in this country. Here, for one thing, the Revolution was delayed, and when it began it was abrupt, thorough, and dramatic. During a decisive phase of that transition, our first significant literary gen-

Leo Marx, "The Machine in the Garden," *The New England Quarterly*, 29 (March 1956), 27–41. Copyright © 1956, *The New England Quarterly*. Reprinted without footnotes by permission.

eration, that of Hawthorne and Emerson, came to maturity. Hence it may be said that our literature, virtually from the beginning, has embodied the experience of a people crossing the line which sets off the era of machine production from the rest of human history. As Emerson said, speaking of the century as the "age of tools," so many inventions had been added in his time that life seemed "almost made over new." This essay demonstrates one of the ways in which a sense of the transformation of life by the machine has contributed to the temper of our literature. The emphasis is upon the years before 1860, because the themes and images with which our major writers then responded to the onset of the Machine Age have provided us with a continuing source of meaning.

Some will justifiably object, however, that very little of the work of Emerson, Thoreau, Hawthorne, or Melville actually was *about* the Industrial Revolution. But this fact hardly disposes of the inquiry; indeed, one appeal of the subject is precisely the need to meet this objection. For among the many arid notions which have beset inquiries into the relations between literature and society, perhaps the most barren has been the assumption that artists respond to history chiefly by making history their manifest subject. As if one might adequately gauge the imaginative impact of atomic power by seeking out direct allusions to it in recent literature. Our historical scholars do not sufficiently distinguish between the setting of a literary work (it may be institutional, geographical, or historical) and its subject-matter or theme. A poem set in a factory need be no more about industrialism than *Hamlet* about living in castles. Not that the setting is without significance. But the first obligation of the scholar, like any other reader of literature, is to know what the work is about. Only then may he proceed to his special business of elucidating the relevance of the theme to the experience of the age.

But here a difficulty arises: the theme itself cannot be said to "belong". to the age. It is centuries old. The Promethean theme, for example, belongs to no single time or place; history periodically renews man's sense of the perils attendant upon the conquest of nature. This obvious fact lends force to the view, tacit postulate of much recent criticism, that what we value in art derives from (and resides in) a realm beyond time, or, for that matter, society. Yet because the scholar grants his inability to account for the genesis of themes, he need not entertain a denial of history. True, he should not speak, for example, of *the* literature of industrialism, as if there were serious works whose controlling insights originate in a single, specific historical setting. But he has every reason to assume that certain themes and conventions, though they derive from the remote past, may have had a peculiar relevance to an age suddenly aware that machines were making life over new. That, in any case, is what seems to have happened in the age of Emerson and Melville. Because our writers seldom employed industrial settings until late in the century we have thought that only then did the prospect of a mechanized America affect their vision of life. My view is that an awareness of the Machine Revolution has been vital to our literature since the 1830s.

But what of a man like Hawthorne, whom we still regard as the "pure" artist, and whose work apparently bears little relation to the Industrial Revolution of his age? In his case it is necessary to demonstrate the importance for his work of matters about which he wrote virtually nothing. Here is "Ethan Brand," a characteristic story developed from an idea Hawthorne recorded in 1843: "The search of an investigator for the Unpardonable Sin;—he at last finds it in his own heart and practice." The theme manifestly has nothing to do with industrialization. On the contrary, it is traditional; we correctly associate it with the Faust myth. Nevertheless, some facts about the genesis of the tale are suggestive. For its physical details, including characters and landscape, Hawthorne drew upon notes he had made during a Berkshire vacation in 1838. At that time several new factories were in operation along the mountain streams near North Adams. He was struck by the sight of machinery in the green hills; he took elaborate notes, and conceived the idea of a malignant steam engine which attacked and killed its human attendants. But he did nothing with that idea, or with any of his other observations upon the industrialization of the Berkshires. And the fact remains that nowhere, in "Ethan Brand" or the notebooks, do we find any explicit evidence of a direct link between Hawthorne's awareness of the new power and this fable of the quest for knowledge of absolute evil.

II

Nevertheless this connection can, I believe, be established. What enables us to establish it is the discovery of a body of imagery through which the age repeatedly expressed its response to the Industrial Revolution. This "imagery of technology" is decisively present in "Ethan Brand."

Although the theme is traditional, some components of the tale take on their full significance only when we consider what was happening in America at the time. Between 1830 and 1860 the image of the machine, and the idea of a society founded upon machine power suddenly took hold of the public imagination. In the magazines, for example, images of industrialism, and particularly images associated with the power of steam, were widely employed as emblems of America's future. They stood for progress, productivity, and, above all, man's new power over nature. And they invariably carried a sense of violent break with the past. Later, looking back at those years, Henry Adams compressed the essential feeling into his account of the way "he and his eighteenth century . . . were suddenly cut apart—separated forever . . ." by the railroad, steamship, and telegraph. It is the suddenness and finality of change—the recent past all at once a green colonial memory—to which American writers have persistently called attention. The motif recurs in our literature from *Walden* to *The Bear*.

But at first our writers did not respond by writing *about* the Industrial Revolution. Long before they knew enough to find concepts for the experience, as Adams later could do, they had invested their work with ideas and emotions it provoked. To this end they drew upon images of technology already familiar to the public. Such collective representations, or "cultural images," allowed them to express what they could not yet fully understand. And at times they heightened these images to the intensity of symbols. The symbolism of Hawthorne and Melville was, after all, designed to get at circumstances which gave rise to conflicting emotions, and which exceeded, in their complexity, the capacities of understanding. Indeed, there is reason to believe that the unprecedented changes then taking place may have provided a direct impetus to the use of symbolic techniques. Hawthorne admitted as much in explaining why he required the image of the railroad to convey that sense of loneliness in the crowd he thought characteristic of the new America. This image, he said, enabled him to present the feeling of a "whole world, both morally and physically, . . . detached from its old standfasts and set in rapid motion."

Now, as the statements of Emerson, Adams, and Hawthorne suggest, the evocative power of the imagery of industrialism is not to be attributed to any intrinsic feature of machines. What gives rise to the emotion is not the machine, but rather its presence against the felt background of the older historic landscape. The American landscape, in fact, accounts for another singular feature of the response to our Industrial Revolution. In this country mechanization had been arrested, among other things, by space—the sheer extent of the land itself. Then, suddenly, with the application of the energy of heat to transport, this obstacle had been overcome. Hence the dramatic decisiveness of the changes in Hawthorne's time, when steam power was suddenly joined to the forces already pressing to occupy the virgin land. In America machines were preëminently conquerors of nature—nature conceived as space. They blazed across a raw landscape of wilderness and farm.

Now it is hardly necessary to discuss the high value, esthetic, moral, and even political, with which the landscape had so recently been invested. Henry Nash Smith has indicated how central a place images of the landscape occupied in the popular vision of America's future as new Garden of the World. The sudden appearance of the Machine in the Garden deeply stirred an age already sensitive to the conflict between civilization and nature. This symbolic tableau recorded the tension between two opposed conceptions of man's relations with his universe. The society prefigured by the myth of the Garden would celebrate a passive accommodation to nature's law. There, survival would depend upon organic production or growth. But, on the other hand, the machine foretold an economy designed by man's brain, and it implied an active, indeed proud, assertion of his dominion over nature. Hence the writers of the American "renaissance," not unlike Shakespeare's contemporaries, confronted one of the more rewarding situations history bestows upon art: the simultaneous attraction of two visions of a people's destiny, each embodying a discrete view of human experience, and each, moreover, accompanied by fresh and vivid symbols. The theme was as old as the story of Prometheus and Epimetheus, but its renewed vitality in Hawthorne's day may be attributed to the power with which fire was making life over new.

III

In 1838, five years before Hawthorne had formulated the moral germ of "Ethan Brand," he had been struck by an actual sight of this change in American society. "And taking a turn in the road," he wrote, "behold these factories.... And perhaps the wild scenery is all around the very site of the factory and mingles its impression strangely with those opposite ones."

Here was history made visible. What most impressed Hawthorne was a "sort of picturesqueness in finding these factories, supremely artificial establishments, in the midst of such wild scenery." Nevertheless, ten years later, when Hawthorne so thoroughly mined this Berkshire notebook for "Ethan Brand," he passed over these impressions. The factories do not appear in the story. Nor is there any over allusion to industrialization. To speculate about the reasons for this "omission" would take us far afield. Whatever the reason, the important fact is not that "Ethan Brand" contains no mention of the factories themselves, but that the ideas and emotions they suggested to Hawthorne are central to the story.

A sense of loss, anxiety, and dislocation hangs over the world of "Ethan Brand." The mood is located in the landscape. At the outset we hear that the countryside is filled with "relics of antiquity." What caused this melancholy situation becomes apparent when Hawthorne describes the people of the region. Brand has returned from his quest. Word is sent to the village, and a crowd climbs the mountain to hear his tale. From among them Hawthorne singles out several men: the stage-agent, recently deprived of his vocation; an old-fashioned country doctor, his useful days gone by; and a man who has lost a hand (emblem of craftsmanship?) in the "devilish grip of a steam-engine." He is now a "fragment of a human being." Like the Wandering Jew and the forlorn old man who searches for his lost daughter (said to have been a victim of Ethan's experimental bent), all are derelicts. They are victims of the fires of change. Like the monomaniac hero himself, all suffer a sense of not belonging.

This intense feeling of "unrelatedness" to nature and society has often been ascribed to the very historical forces which Hawthorne had observed in 1838. Discussing the intellectual climate of that era, Emerson once remarked that young men then had been born with knives in their brains. This condition was a result, he said, of the pervasive "war between intellect and affection." He called it "detachment," and found it reflected everywhere in the age: in Kant, Goethe's *Faust*, and in the consequences of the new capitalist power. "Instead of the social existence which all shared," he wrote, "was now separation." Whatever we choose to call it— "detachment" or "alienation' (Karl Marx), or "anomie" (Emile Durkheim) or "dissociation of sensibility" (T. S. Eliot)—this is the malaise from which Ethan suffered. Though there are important differences of emphasis, each of these terms refers to the state of mind of an individual cut off from a realm of experience said to be an indispensable source of lifes meaning. The source may vary, but it is significant that the responsible agent, or *separator*, so to speak, is invariably identified with science or industrial technology. In this sense Hawthorne's major theme was as vividly contemporary as it was traditional. He gave us the classic American account of the anguish of detachment.

The knife in Ethan's brain was a "cold philosophical curiosity" which led to a "separation of the intellect from the heart." Now it is of the utmost significance that this scientific obsession is said to have literally emanated from the fire. There was a legend about Ethan's having been accustomed "to evoke a fiend from the hot furnace." Together they spent many nights by the fire evolving the idea of the quest. But the fiend always retreated through the "iron door" of the kiln with the first glimmer of sunlight. Here we discover how Hawthorne's earlier impressions of industrialization have been transmuted in the creative process. Here is the conduit through which thought and emotion flow to the work from the artist's experience of his age. In this case the symbolic contrast between fire and sun serves the purpose. It blends a traditional convention (we think of Milton's Hell) and immediate experience; it provides the symbolic frame for the entire story. "Ethan Brand" begins at sundown and ends at dawn. During the long night the action centers upon the kiln or "furnace" which replaces the sun as the source of warmth, light, and (indirectly) sustenance. The fire in the kiln is at once the symbolic source of evil and of the energy necessary to make nature's raw materials useful to man. Moreover, it can be shown that the very words and phrases used to describe this fire are used elsewhere, by Hawthorne, in direct reference to industrialization. In the magazines of the day fire was repeatedly identified with the new machine power. Hence fire, whatever its traditional connotations, is here an emblem, or fragment of an emblem, of the nascent industrial order. The new America was being forged by fire.

But if fire cripples men and devastates the landscape in "Ethan Brand," the sun finally dispels anxiety and evil, restoring man's solidarity with nature. When Ethan dies, his body burned to a brand by the satanic flames which had possessed his soul, the fire goes out and the ravaged landscape disappears. In its stead we see a golden vision of the self-contained New England village. The sun is just coming up. The hills swell gently about the town, as if resting "peacefully in the hollow of the great hand of Providence." In pointed contrast to the murky atmosphere of Ethan's Walpurgisnacht, there is no smoke anywhere. The sun allows perfect clarity of perception. Every house is "distinctly visible." At the center of this pastoral tableau the spires of the churches catch the first rays of the sun. Now the countryside is invested with all the order and serenity and permanence which the fire had banished. This harmony between man and nature is then projected beyond time in the vision of a stepladder of clouds on which it seemed that (from such a social order?) "mortal man might thus ascend into heavenly regions." Finally, though he had already hinted that stage coaches were obsolete, Hawthorne introduces one into this eighteenth-century New England version of the Garden of Eden.

Beneath the surface of "Ethan Brand" we thus find many of the ideas and emotions aroused by the Machine's sudden entrance into the Garden. But this is not to say that the story is *about* industrialization. It is about the consequences of breaking the magic chain of humanity. That is the manifest theme and, like the symbols through which it is developed, the theme is traditional. His apprehension of the tradition permits Hawthorne to discover meanings in contemporary facts. On the other hand, the capacity of this theme and these images to excite the imagination must also be ascribed to their vivid relevance to life in modern America. This story, in short, is an amalgam of tradition, which supplies the theme, and experience, which presents the occasion, and imagery common to both.

IV

But it may be said that, after all, this is merely one short story. The fact remains, however, that the same, or related, images may be traced outward from this story to Hawthorne's other work, the work of his contemporaries, the work of many later writers, and the society at large.

It is revealing, for example, that Hawthorne so often described his villains as alchemists, thereby associating them with fire and smoke. We recall that Aylmer, the scientist in "The Birthmark," made a point of building his laboratory underground to avoid sunlight. Or consider Rappaccini, whose experiments perverted the Garden itself. His flowers were evil because of their "artificialness indicating . . . the production . . . no longer of God's making, but the monstrous offspring of man's depraved fancy. . . ." From Hamilton's "Report on Manufactures" in 1791 until today, American thinking about industrialism, in and out of literature, has been tangled in the invidious distinction between "artificial' and "natural" production. These adjectives, like so much of American political rhetoric, along with Hawthorne's theme of isolation, are a characteristic legacy of agrarian experience. They are expressions of our native tradition of pastoral with its glorification of the Garden and its consequent identification of science and technology with evil. To Hawthorne's contemporaries, Emerson, Thoreau, and Whitman, the sun also represented the primal source of redemption. "The sun rose clear," Thoreau tells us at the beginning of *Walden*; though he notes that the smoke of the train momentarily obscures its rays, the book ends with a passionate affirmation of the possibility of renewed access, as in "Ethan Brand," to its redeeming light: "The sun is but a morning star."

In *Moby-Dick*, published three years after "Ethan Brand," the identical motif emerges as a controlling element of tragedy. The "Try-Works," a crucial chapter in Ishmael's progressive renunciation of Ahab's quest, is quite literally constructed out of the symbols of "Ethan Brand." Again it is night, and vision is limited to the lurid light of a kiln or furnace. Fire again is a means of production, rendering the whale's fat, and again it is also the source of alienation. Ishmael, at the helm, controls the ship's fate. Like Ethan he momentarily succumbs to the enchantment of fire, and so nearly fulfills Ahab's destructive destiny. But he recovers his sanity in time, and tells us: "Look not too long in the face of the fire, O Man! . . . believe not the *artificial fire* when its redness makes all things look ghastly. Tomorrow, in the *natural sun*, the skies will be bright; those who glared like devils in the forking flames, the more will show in far other, at least gentler relief; the glorious, golden, glad sun, the only true lamp—all others but liars!"

From this passage we may trace lines of iconological continuity to the heart of Melville's meaning. When the *Pequod* sailed both Ahab and Ishmael suffered the pain of "detachment." But if the voyage merely reinforced Ahab's worship of the power of fire, it provoked in Ishmael a reaffirmation of the Garden. Ahab again and again expressed his aspiration in images of fire and iron, cogs and wheels, automata, and manufactured men. He had his "humanities," and at times was tempted by thoughts of "green land," but Ahab could not finally renounce the chase. In *Moby-Dick* space is the sea—a sea repeatedly depicted in images of the American landscape. The conquest of the whale was a type of our fated conquest of nature itself. But in the end Ishmael in effect renounced the fiery quest. He was cured and saved. His rediscovery of that pastoral accommodation to the mystery of growth and fertility was as vital to his salvation as it had been to the myth of the Garden. The close identity of the great democratic God and the God of the Garden was a central facet of Melville's apocalyptic insight.

His was also a tragic insight. Ahab and Ishmael, representing irreconcilable conceptions of America's destiny, as indeed of all human experience, were equally incapable of saving the *Pequod*. From Melville to Faulkner our writers have provided a desperate recognition of this truth: of the attributes neccessary for survival, the Ahabs alone have been endowed with the power, and the Ishmaels with the perception. Ishmael was saved. But like one of Job's messengers, he was saved to warn us of greater disasters in store for worshippers of fire. In this way imagery associated with the Machine's entrance into the Garden has served to join native experience and inherited wisdom.

Science and Technology

CARL W. CONDIT

The skyscrapers that dominate the urban landscape are symbols of modern civilization. Like the western steamboat and the "American System" of manufacturing, the skyscraper represents a distinctively American contribution to technology. In contrast to the alienation of intellectuals like Hawthorne, the engineers and architects who built great bridges and skyscrapers experienced an exultant sense of power as a result of the triumphs of technology. Their sense of power was enhanced by a profound transformation in technology in the latter nineteenth century. Technology broke with centuries-old craft traditions and found new foundations in science, which greatly increased the power of technology and accelerated the tempo of change. But, unfortunately, technological progress does not automatically lead to human betterment. The skyscrapers contributed to congestion and other urban problems. Some people think they were human disasters. Carl Condit is a professor of art at Northwestern University and the author of *The Chicago School of Architecture* (Chicago, 1964) and *American Building* (Chicago, 1968).

It is now a matter of common consent that Louis Sullivan (1856–1924) was the first great modern architect, the first to create a new and powerful vocabulary of forms derived from the major cultural determinants of his age. He was the most imaginative and the most articulate figure among a small group of creative men in Europe and America who, suddenly around 1890, struck out in a new direction with the deliberate intention of breaking once and for all with the traditional architectural forms of the classical

Carl W. Condit, "Sullivan's Skyscrapers as the Expressions of Nineteenth Century Technology," *Technology and Culture*, 1 (Winter 1959), 78–93. Copyright © 1959, Society for the History of Technology. Reprinted without footnotes by permission of the author and the publisher, The University of Chicago Press.

and medieval heritage. In Europe the movement called itself *Art Nouveau,* its initiator being the Belgian architect Baron Victor Horta (1861–1947). In the United States it was at first confined largely to Chicago, where the fire of 1871 prepared the way for one of the most exuberant outbursts of creative activity in nineteenth-century architecture. The leadership of this movement, now known as the Chicago school, was initially in the hands of William Le Baron Jenney (1832–1906), but by 1890 it had passed to Sullivan and his engineering partner, Dankmar Adler (1844–1900). Within a single decade Adler and Sullivan moved rapidly, if irregularly, from close dependence on past architectural styles to an organic form which derived its character from the industrial and scientific culture which had swept everything before it in the Western world.

By the last decade of the century Sullivan had developed in preliminary terms his organic theory of building art, a system which was later to be presented at length in his major writings, *Kindergarten Chats* (1901–1902) and *The Autobiography of an Idea* (1922). The philosophy of architecture offered in these works contains extensive ethical and social elements as well as formal and aesthetic. Since the doctrine has been discussed, analyzed, and interpreted in detail by historians and critics, we need not here inquire into it at length. Our purpose is to find, if we can, the broad symbolic meaning of Sullivan's major works, for which it may be useful to summarize some of the fundamental ideas in his system of thought.

Sullivan's interest in structural engineering—in part, of course, the product of professional necessity—early developed into a wide-ranging enthusiasm for science as a whole. It centered mainly in biology, from which his organic theory in part stemmed, but it included the new physical theories as well. He read Darwin, Huxley, Spencer, and Tyndall at length and was well acquainted with the writers who were then developing the seminal theories of building art in the past century, chiefly Ruskin, Morris, and Viollet-le-Duc. What distinguishes Sullivan's thought is his profound grasp of the social basis, the responsibility, and the problem of the arts in a technical and industrial society. He felt that he had discovered the rule with no exceptions (to use his own phrase) in the concept "form follows function," but the key to his philosophy lies in the proper understanding of the word *function.* An organic architecture, he believed, is one which grows naturally or organically out of the social and technical factors among which the architect lives and with which he must work.

These factors embrace not only the technical and utilitarian problems of building but also the aspirations, ideals, and needs, both material and psychological, of mankind. Thus *functionalism* involved for him something much wider and deeper than utilitarian and structural considerations, as important as these are.

To Sullivan the creation of a genuine architectural style was not a matter of historical styles or of dipping into a vocabulary of contemporary forms and details in order to secure a style which the architect might feel to be consonant with the life of this time. The architect must first recognize the importance of true aesthetic expression for the symbolic recreation, the harmonization, and the emotional enrichment of the many practical and intellectual elements of contemporary civilization. In European and American society at the end of the nineteenth century such an art would begin, by necessity, with the fundamentals: industry, technology, and science. It is the task of the architect, as Sullivan conceived it, to take the products of techniques, on the one hand, and the logic and order of a scientific technology, on the other, and mold them into a form uniting both in a single aesthetic expression. An architecture so developed means the humanization through aesthetic statement of the often cold and nonhuman facts of industrial techniques.

The early application of this complex philosophy of the organic to a specific building problem appeared in a document which has become a classic of modern theory, "The Tall Office Building Artistically Considered," first published in *Lippincott's Magazine* in 1896. Scattered throughout *The Autobiography of an Idea*, which is Sullivan's final testament, are many sentences of an epigrammatic character that summarize his thought; for example, "As the people are within, so the buildings are without," or again, "It is the task of the architect to build, to express the life of his own people."

The realization of this program in actual commissions reached its mature form in the four largest and most impressive buildings which Adler and Sullivan designed. The first is the Auditorium Building, now Roosevelt University in Chicago, designed in 1886–1887 and opened in 1889. The design of this great building, with its huge masonry bearing walls, was much influenced by Richardson (1838–86) and the Romanesque-like forms which he handled so brilliantly. But it marks a transition toward the open and dynamic wall forms that were soon to become the distinguishing feature of Sullivan's work. In the year following the completion of the

Auditorium he struck out in a new direction to produce one of the most remarkable exhibitions of sheer architectural originality in his own or any age, the steel-framed Wainwright Building in St. Louis (1890–91). A few years later the formal character of this structure was refined and enriched in the Guaranty, later Prudential, Building in Buffalo (1894–95). In the last of his large commissions, before the poverty and neglect of his later years, he turned in still another direction and produced a radically different kind of expression in the Carson Pirie Scott Store in Chicago, built in three parts over the years from 1899 to 1906, although designed as early as 1896.

Behind these steel-framed buildings lay a long preparation in the history of iron construction. The use of iron as a structural material goes back to classical antiquity, but it did not appear wholly emancipated from masonry until the construction of Darby and Pritchard's cast iron arch over the River Severn at Coalbrookdale, England, in 1775–79. The first building with interior columns and beams of iron was William Strutt's Calico Mill at Derby (1793). It was thirty-five years before iron members appeared in American buildings and 1850 before complete iron construction was established, largely through the work of the New York inventors and builders Daniel Badger and James Bogardus. The remainder of the century saw steady progress in the techniques of cast and wrought iron and later steel framing, reaching its culmination in the skyscrapers of New York and Chicago, in which all the essential features of the modern commercial building were given a practical demonstration.

Thus, by 1890, the technical means of a new building art were available to Sullivan. It remained for him to transmute the structural solutions to these unprecedented functional requirements into a symbolic art. His organic philosophy had already come to exist at least in an inchoate form, but so broad an approach to architectural design could not lead directly to a specific kind of formal expression. The key to the process of transmutation may be found, I think, in those passages of Sullivan's writings in which he gives voice to his feelings about particular architectural and structural achievements of his age. We have already mentioned his debt to Richardson. There is a chapter in *Kindergarten Chats*—Number VI, "The Oasis"—in which he acknowledges this debt, and it is the first of various passages that lead us to an understanding of Sullivan's inner purpose. He describes for us, in ironic and impressionistic metaphors, his strong emotional reaction to Richardson's Marshall Field Wholesale Store in Chicago (1885–87) and to what it stands for.

> Let us pause, my son, at this oasis in our desert. Let us rest awhile beneath its cool and satisfying calm, and drink a little at this wayside spring. . . .
> You mean, I suppose, that here is a good piece of architecture for me to look at—and I quite agree with you.
> No; I mean here is a *man* for you to look at. A man that walks on two legs instead of four, has active muscles, heart, lungs, and other viscera; a man that lives and breathes, that has red blood; a real man, a manly man; a virile force—broad, vigorous and with a whelm of energy—an entire male.
> I mean that stone and mortar, here, spring into life, and are no more material and sordid things, but, as it were, become the very diapason of a mind rich-stored with harmony. . . .
> Four square and brown, it stands, in physical fact, a monument to trade, to the organized commercial spirit, to the power and progress of the age, to the strength and resource of individuality and force of character; spiritually, it stands as the index of a mind, large enough, courageous enough to cope with these things, master them, absorb them and give them forth again, impressed with the stamp of large and forceful personality; artistically, it stands as the creation of one who knows well how to choose his words, who has somewhat to say and says it—and says it as the outpouring of a copious, direct, large and simple mind.

It is clear that Sullivan was profoundly moved by Richardson's building, but even a fine work of architecture did not arouse in him the powerful emotions that were evoked by the great achievements of the bridge engineers in the nineteenth century. There are several illuminating passages in *The Autobiography*, among the most remarkable in the book, in which he tries to analyze his emotional and philosophic response to these monuments of pure structural form. The first records a childhood experience in which he saw a chain suspension bridge over the Merrimack River (possibly Finley's Bridge of 1810, near Newburyport, Massachusetts). The description is loaded with the most extreme expressions of feeling.

> Mechanically he ascended a hill . . . musing, as he went, upon the great river Merrimac. . . . Meanwhile something large, something dark was approaching unperceived; something ominous, something sinister that

silently aroused him to a sense of its presence. . . . The dark thing came ever nearer, nearer in the stillness, became broader, looming, and then it changed itself into full view—an enormous terrifying mass that overhung the broad river from bank to bank. . . .

He saw great iron chains hanging in the air. How could iron chains hang in the air? He thought of Julia's fairy tales and what giants did. . . . And then he saw a long flat thing under the chains; and this thing too seemed to float in the air; and then he saw two great stone towers taller than the trees. Could these be giants? . . . [A page follows in which Sullivan records how he ran frightened to his father to tell him that the giants would eat him.]

So [his father] explained that the roadway of the bridge was just like any other road, only it was held up over the river by the big iron chains; that the big iron chains did not float in the air but were held up by the stone towers over the top of which they passed and were anchored firmly into the ground at each end beyond the towers; that the road-bed was hung to the chains so it would not fall into the river. . . . On their way to rejoin Mama, the child turned backward to gaze in awe and love upon the great suspension bridge. There, again, it hung in the air—beautiful in power. The sweep of the chains so lovely, the roadway barely touching the banks. And to think it was made by men! How great must men be, how wonderful; how powerful, that they could make such a bridge; and again he worshipped the worker.

In later years, on his way to becoming an established architect in partnership with one of the great building engineers of his time, Sullivan came to understand how these miracles were accomplished. Then he was prepared to pay his fullest tribute to the bridge engineers and to record it again in his *Autobiography*.

About this time two great engineering works were under way. One, the triple arch bridge to cross the Mississippi at St. Louis, Capt. Eades [*sic*], chief engineer; the other, the great cantilever bridge which was to cross the chasm of the Kentucky River, C. Shaler Smith, chief engineer, destined for the use of the Cincinnati Southern Railroad. In these two growing structures Louis's soul became immersed. In them he lived. Were they not his bridges? Surely they were his bridges. In the pages of the *Railway Gazette* he saw them born, he watched them grow. Week by week he grew with them. Here was Romance, here again was man, the great adventurer, daring to think, daring to have faith, daring to do. Here again was to be set forth to view man in his power to create beneficently.

Here were two ideas differing in kind. Each was emerging from a brain, each was to find realization. One bridge was to cross a great river, to form the portal of a great city, to be sensational and architectonic. The other was to take form in the wilderness, and abide there; a work of science without concession. Louis followed every detail of design, every measurement; every operation as the two works progressed from the sinking of the caissons in the bed of the Mississippi, and the start in the wild of the initial cantilevers from the face of the cliff. He followed each, with the intensity of personal identification, to the finale of each. Every difficulty he encountered he felt to be his own; every expedient, every device, he shared in. The chief engineers became his heroes; they loomed above other men. The positive quality of their minds agreed with the aggressive quality of his own. In childhood his idols had been the big strong men who *did* things. Later on he had begun to feel the greater power of men who could *think* things; later the expansive power of men who could *imagine* things; and at last he began to recognize as dominant the will of the Creative Dreamer: he who possessed the power of vision needed to harness Imagination, to harness the intellect, to make science do his will, to make the emotions serve him—for without emotion—nothing.

There is a distinct strain of romanticism in this passionate devotion to the builder, perhaps even a Nietzschean quality in the worship of creative power. For Sullivan came to see in science and technology the triumphant assertion of man's will expressing itself in a wholly new way. As he himself put it, "Louis saw power everywhere; and as he grew on through his boyhood, and through the passage to manhood, and to manhood itself, he began to see the powers of nature and the powers of man coalesce in his vision into an IDEA *of power*. Then and only then he became aware that this idea was a *new idea*,—a complete reversal and inversion of the commonly accepted intellectual and theological concept of the nature of man."

Thus Sullivan conceived of a bridge as the personal testament of a man, a testament expressing a unification of the highest energies and skills of the age. What distinguishes these achievements is not only the technical virtuosity that men like Eads and Smith commanded, but the integration of many streams of technical and scientific progress in the nineteenth century. For it was the age that saw the transformation of building from an empirical and pragmatic technique into an exact science. Since Sullivan chose the St. Louis and Dixville bridges as his examples, we may use them as representatives of the transformation that made possible their design and construction. Eads Bridge (1868–74) is the earlier of the two, and so we may begin with an analysis which reveals how such structures brought to focus the various scientific and technical currents.

The general staff of Eads Bridge consisted of James B. Eads as chief engineer, Charles Pfeiffer and Henry Flad as principal assistants, and William Chauvenet, Chancellor of Washington University, as mathematical consultant. The choice of Eads, who had never built a bridge before, as head of the St. Louis project rested in large part on his intimate knowledge of the river. For the builder who proposed to found his piers on the rock far below its bed, it was a formidable obstacle indeed. The pilots could read its surface with remarkable skill for the hidden snags and bars that once menaced them, but only Eads knew at first hand its fluid, shifting, treacherous bed. By means of the diving bell which he had invented, he was able to investigate the bottom directly, and he had seen its depth change from 20 to 100 feet at obstacles in the bed as the result of the scouring action of currents. For the first time the topographic and geological surveys of the bridge site could be carried on to a certain extent under water.

With the design of his bridge substantially completed and sufficient capital available, Eads began clearing the site and constructing caissons in the summer of 1867, but difficulties with his iron and steel contractor soon required a suspension of operations. Eads had already decided to substitute steel for the traditional cast iron in the arches and thus became the first to introduce the stronger metal into American building techniques. The Carnegie-Kloman Company at first found it impossible to roll pieces with the physical properties that Eads demanded. The earlier cast steel samples had already failed in the testing machines. At this point Eads insisted on the costly and hitherto unused chrome-steel, an innovation which was to have wide implications for structural and mechanical engineering. Equally important was the application of the methods of experimental science to the investigation of the physical properties of the metal. Eads Bridge is the first major structure in the United States in which testing machines, which had been developed over the previous thirty years in Europe, played a vital role in the successful completion of the bridge.

The initial problem solved, construction was resumed in 1868. Eads began with the east, or Illinois, pier, where the maximum depth of bedrock offered the most serious challenge. The pneumatic caisson was a necessity, and thus Eads became the first to introduce its use in the United

States, anticipating Roebling by a year. It had been used in Europe since 1849, when Lewis Cubitt and John Wright developed it for the construction of the piers of a span at Rochester, England. Eads built a cylindrical iron-shod caisson of massive timbers heavily reinforced with iron bands. Its diameter was 75 feet, the working chamber 8 feet deep. Within this huge enclosure the masonry pier was built up, the weight of the masonry forcing the cutting edge into the river bed. The Eads caisson extended continuously up to the water surface, successive rings being added as it sank lower. The caisson for the Illinois pier reached bedrock at 123 feet below water level at the time of construction. Five months of excavation and pumping were required to uncover the foundation rock. Since the top stratum of the bedrock rises steadily from the east to the west bank, the caissons for the center and west piers had to be sunk to a progressively smaller depth, reaching a minimum of 86 feet at the St. Louis pier. The river piers and the masonry arches of the west approach were completed in 1873.

The construction of the steel arches and the wrought iron superstructure was a relatively simple matter after the dangerous work on the piers and required only about one-fifth of the time. In this part of the project Eads introduced another of his important innovations. The tubular arches were erected without falsework by the method of cantilevering them out from the piers to the center of the span. All arches were built out simultaneously from their piers so that the weights of the various cantilevers would balance each other and on completion the horizontal thrusts of adjacent arches would cancel each other. With the arches in place, the spandrel posts and the two decks were erected upon them. The bridge was completed and opened to traffic in 1874. The finished structure between abutments is divided into three spans, the one at the center 520 feet long, the two at the sides 502 feet each, the rise for all of them 45 feet. Eight tubular arches, four for each deck, constitute the primary structure of each span. Wrought-iron spandrel posts and transverse bracing transmit the load of the two decks to the arches. The upper one carries a roadway, the lower a double-track railroad line.

The arches of Eads Bridge are the hingeless or fixed-end type and hence are statically indeterminate structures. It is possible that Chauvenet was familiar with the recent work of French theorists in the solution of problems arising from arches of this kind, and certainly he knew of the many carefully designed wrought iron arches which had been built by

French engineers before 1865. The successful attack on problems of indeterminacy was the product of an international effort in which a great many mathematicians had a hand, among them the great English physicist James Clerk Maxwell. The chief figure in the development of methods of stress analysis for fixed and two-hinged arches was Jacques Antoine Bresse, the first edition of whose *Applied Mechanics* was published at Paris in 1859. But for all the mathematical computations of Eads and Chauvenet, they relied to a great extent—as the engineers always did until the last decade of the century—on empirical approximations and gross overbuilding. Eads calculated that his bridge would be capable of sustaining a total load of 28,972 tons uniformly distributed—about four times the maximum that can be placed upon it—and of withstanding the force of any flood, ice jam, or tornado that the Mississippi Valley could level against it. Now in its ninth decade of active service, the bridge carries a heavy traffic of trucks, busses, automobiles, and the freight trains of the Terminal Railroad of St. Louis.

Sullivan could hardly have chosen a better example to represent the new power of his age. As a work of structural art Eads Bridge remains a classic. In its method of construction and its material, in the testing of full-sized samples of all structural members and connections, in the thoroughness and precision of its technical and formal design, and in the close association of manufacturer and builder, it stands as a superb monument to the building art. It is, moreover, an architectural as well as an engineering achievement. Eads was careful to reduce his masonry elements to the simplest possible form, depending on the rich texture of the granite facing to provide the dignity and sense of restrained power that he was consciously striving for. Nowhere does the masonry extend above the line of the parapet to distract attention from the overall profile, the major parts, and their relation to each other. The tight curve of the arches is the primary visual as well as structural element, and Eads knew that the best he could do was to give full expression to the combination of stability and energy implicit in the form.

The Dixville Bridge posed an entirely different and somewhat less formidable problem. The solution, moreover, belongs strictly to the nineteenth century, the particular truss form employed having been abandoned before the beginning of the twentieth century. It was the decision to use the cantilever principle for a large railway bridge, for which there was only the slightest of precedents, that excited Sullivan's interest. The occasion in this case was the necessity of bridging the Kentucky River at Dixville, Kentucky, for the Cincinnati Southern Railway. The engineer in charge of the project was L. F. G. Bouscaren (1840–1904), chief engineer of the railroad company and designer of its Ohio River Bridge at Cincinnati, a structure which was built simultaneously with the Dixville Bridge and which contained the longest simple truss span in the world at the time of its erection. There is no question that Bouscaren deserves as much credit as Charles Shaler Smith for the Dixville project, but Sullivan and the rest of posterity have always honored Smith and forgotten the other half of the team.

The chief problem at Dixville was that of erecting the trusses. The Kentucky River gorge at this point is 1,200 feet wide and 275 feet deep. The river has always been subject to flash floods of disastrous proportions, a maximum rise of 40 feet in one day having already been recorded when the two engineers made their preliminary survey. The use of falsework under such conditions was out of the question. Smith originally planned to build a continuous Whipple-Murphy deck truss, 1,125 feet long extending over three spans of 375 feet each. At this point Bouscaren made the proposal that hinges be introduced into the truss at two points, one at each end of the bridge between the shore and the nearest pier. The use of hinges in a continuous beam or truss was first proposed by the German engineer and theorist Karl Culmann in his *Graphical Statics* (1866). The introduction of hinges and the resulting transformation of the parts of tne beam on either side of the support into cantilevers has the consequence that the action of the member more nearly conforms to the theoretical curve of stress distribution, or stress trajectory, as it is sometimes called. The first large cantilever bridge whose design seems clearly to have beer. influenced by Culmann's theory was Heinrich Gerber's bridge over the Main River at Hassfurt, Germany (1867). The structure excited wide interest and was undoubtedly known to Smith and Bouscaren.

Several other factors, however, led to Bouscaren's decision. In addition to improvement in the efficiency of the truss action, the engineers were concerned to prevent the excessively high stresses which would have occurred in the continuous truss as a consequence of pier settlement. Further, the successful construction of Eads Bridge by the method of cantilevering the arches out from the abutments and piers suggested not only a similar mode of construction for continuous trusses but also the possibility of using the cantilever as a permanent structural form. By adopting Bouscaren's suggestion the designers turned the bridge into a combination of types which were, in succession from shore to shore, a semi-floating span fixed at the abutment and hinged at the free end, a 75-foot cantilever, a simple truss acting as anchor span to the cantilevers, and so on in reverse order to the opposite shore. The material of the structure throughout was wrought iron. There was, as we noted, little precedent for a bridge of this kind, and Smith and Bouscaren staked their reputations on it. They saw it through to successful completion, but the increasing weight of traffic required that it be replaced in 1911 by a steel bridge built on the same masonry.

Sullivan's intuitive grasp of the meaning of these bridges was perfectly sound. The union of science and technology which made them possible was the creation of men who possessed a rare combination of faculties: they were men who could imagine and think things and who, when they translated the products of imagination into physical fact, did so on a heroic scale. It was difficult not to be impressed, however little one understood the methods of their achievement. Sullivan was profoundly moved, and he knew that he would have to create a building art which could give voice to these powerful feelings and thus evoke them in others.

The Auditorium Building marks the initial step, although what is visible both inside and out at first seems to have little to do with the great achievements of the Age of Iron. Perhaps it is the fact of the building itself, rather than what it says in detail, that heralds a new epoch in architectural form. The exterior walls, as magnificent as they are in their architectonic power, are masonry bearing elements disposed in the long-familiar system of stout piers and arcades. Inside this uniform block, however, is the most extraordinary diversity of internal volumes that one can house in a single structure. The huge theatre, seating 4,000 people, is surrounded by a block of offices on the west and south and by a hotel on the east. Offices, hotel rooms, lobbies, small dining rooms, and other utilitarian facilities are carried on a complete system of framing composed of cast-iron columns and wrought-iron beams. The great vault of the theatre is hung from a series of parallel elliptical trusses which are suspended in turn from horizontal trusses immediately under the roof. The same construction, on a smaller scale, supports the vault of the main dining room. Above the theatre at one end still another type of truss is used to support the ceiling of the rehearsal room. The vaulted enclosures are in no way like the traditional barrel vaults of Roman and medieval building. They are great wide-span cylinders of elliptical or segmental sections which were derived from the huge trainsheds of the nineteenth-century railway station and thus constitute a metamorphosis of a quasi-monumental utilitarian form into one element of an aesthetic complex. The system of truss framing in the Auditorium grew out of the inventions of the bridge engineers. Under the once brilliant colors and intricate interweaving patterns of Sullivan's ornament, none of this construction is visible, yet the light screens with their plastic detail, and the multiplicity of shapes and volumes could not have been created without the structural means that Adler employed. As a matter of fact, the Auditorium embraces every basic structural technique available to the nineteenth-century builder.

The three purely commercial buildings—the Wainwright, the Prudential, and the Carson store—rest on the more advanced structural technique of complete steel framing without masonry bearing members of any kind other than the concrete column footings, but because of the uniformity and relative simplicity of their interior spaces they are much less complex in their construction than the Auditorium. Yet it was precisely here that Sullivan saw his opportunity: now he could take full advantage of the steel frame in the treatment of the elevations, the obvious parts of the building that everyone had to see. What he was trying to articulate in the three buildings was not simply structure and utility, which the bridge engineers had done in their wholly empirical forms, but rather his complex psychological response to the structural techniques that the engineers employed so boldly. The idea underlying the Wainwright and Prudential buildings is clearly summed up in the celebrated passage of the *Autobiography* on the skyscaper. "The lofty steel frame makes a powerful appeal to the architectural imagination where there is any. . . . The appeal and the inspiration lie, of course, in the element of loftiness, in the suggestion of slenderness and aspiration, the soaring quality of a thing rising from the earth as a unitary utterance, Dionysian in beauty."

The formal character of the Prudential Building (which is a larger and more refined counterpart of the Wainwright and thus may be taken to represent the essential quality of both buildings) is an organic outgrowth of its utilitarian functionalism, but it is in no way confined by it. Above an open base, designed chiefly for purposes of display, rises a uniform succession of office floors, identical in function and hence appearance, topped by an attic floor which carries heating returns and elevator machinery and whose external treatment provides a transition to the flat slab that terminates the upward motion of the whole block. Elevator shafts, plumbing, and mechanical utilities are concentrated in an inner core. Sustaining all roof, floor, wall, and wind loads is an interior steel frame covered with fireproof tile sheathing. All this constitutes the empirical answer to utilitarian necessity.

Beyond the empirical form, however, are the wholly aesthetic elements that transform structure into symbolic art. The great bay-wide windows of the base are carefully designed not only to reveal structure but to separate it clearly from all subsidiary elements and thus to give forceful utterance to its potentially dramatic quality. The columns which stand out so clearly are the forerunners of Le Corbusier's *pilotis*, now so common in contemporary building that they have become a cliché. Above the open base appears the single most striking feature, the pronounced upward vertical movement achieved by the closely ranked pier-like bands of which every other one clothes the true structural column, the alternate piers being purely formal additions without bearing function.

The basic theme of this light screen is movement, the dynamic transcendence of space and gravitational thrust, qualities Sullivan long before felt in the "floating" chains and roadway of the Merrimack suspension bridge. In a broader sense the theme suggests the underlying energy of a world of process, of evolutionary growth in living things, or the dynamic of the electric field in physics. The bridge, like the building, is not seen by Sullivan as a static thing but as something which leaps over its natural obstacle and thus becomes a living assertion of man's skill operating through his simultaneous dependence upon and command over nature. Again Sullivan's intuition led him into the right path, for this is exactly how the bridge behaves. We can sense this directly in the suspension bridge with its wire cables and suspenders: it seems alive, constantly quivering under its changing load. Although we can neither see it nor feel it, exactly the same thing is occurring in the dense and massive members of the big railroad truss as the internal stress continuously adjusts itself to the moving weight that it sustains. It took a century and a half of painstaking scientific inquiry to discover this hidden and vital activity.

The rich and intricate ornament that covers the two office buildings, an ornament created by Sullivan which died with him, offers a much more difficult problem of interpretation. It is so subjective that it is scarcely possible to find objective experiences that might have led to the feelings out of which it grew. In its complex, somewhat abstract naturalism, it appears to symbolize the biologically organic. While Sullivan sometimes allowed his ornament to flow in uncontrolled and undifferentiated profusion over much of the surface, he was generally careful to observe the limits of architectural ornamentation. By spreading it in low relief over whole elevations, and by confining a particular pattern to the surface of a certain kind of structural member—column, or spandrel beam—he was able to distinguish in a striking way the separate structural surfaces. Thus his ornament enhances the major elements of the structure and further heightens the vivid sense of movement. It also seems to suggest the diversity underlying the unitary organic statement.

In the Carson Pirie Scott Store Sullivan turned to an entirely different kind of expression, one derived from the dominant mode of the work of the Chicago school. Where he used the close vertical pattern in the older buildings, in the department store he opened the main elevations into great cellular screens which exactly express the neutral steel cage behind them. The form was dictated initially by the requirement for maximum natural light in the store, but again, in many subtle ways, he translated the practical functionalism into art. If the theme of the Wainwright and the Prudential is movement, that of the Carson store is power. Here the elaborate interplay of tension and compression, of thrust and counterthrust in the bridge truss is given a heightened and dramatic statement by means so delicate as almost to escape notice—the careful calculation of the depth of the window reveals and the breadth of the terra-cotta envelope on the columns and spandrels, the narrow band of ornament that enframes the window, the even narrower band that extends continuously along each sill and lintel line to give the whole façade a tense, subdued horizontality. The base in its sheath of ornament is an exact reversal of that of the office buildings. In the Carson store it is a weightless screen, glass and opaque covering (cast iron) forming one unbroken plane, making the cellular wall above literally seem to float free of the earth below it.

In the last analysis Sullivan's civic architecture is a celebration of technique, as is most of the contemporary architecture of which he was the foremost pioneer. But Sullivan had carried the expression far beyond the rather sterile geometry that characterizes most building today. If his work seems limited beside the vastly richer symbolism of medieval and baroque architecture, we may at least say that he was responding to the one coherent order that was discernible in the contradictory currents of nineteenth- and twentieth-century culture. In the absence of a cosmos in which man was conceived to be the central figure, the scientific technology on which building increasingly depended became the one sure basis of architectural and civic art. It is Sullivan's achievement to have understood how this basis could be transmuted into an effective and valid artistic statement.

Engineers in Revolt

EDWIN T. LAYTON, JR.

Scientific technologists were excited by the prospect of man's conquest of nature; but they were dismayed to discover that business sometimes used technology to exploit men—especially when some of them became aware that engineers were among the exploited. Engineers raised fundamental questions of social responsibility; they were among the first groups to attempt to redirect technology to serve human values more effectively. In their efforts they were encouraged by Thorstein Veblen, one of America's most original social thinkers. Though Veblen and the engineers failed to change the direction of technology, the questions they raised are still important and still unresolved. Should technologists assume responsibility for their own creations? Can modern technology be made more responsive to human needs? If society needs to be changed, what form should it take? Edwin T. Layton, Jr., is author of *The Revolt of the Engineers, Social Responsibility and the American Engineering Profession* (Cleveland, 1971) and is a professor of the history of science and technology at Case Western Reserve University.

One of the strangest predictions in the history of social theory was that of Thorstein Veblen who concluded that the engineers would constitute the revolutionary class in America. "The chances of anything like a Soviet in America, therefore," he wrote, "are the chances of a Soviet of Technicians." A group less likely to lead a revolution in America would be hard to imagine. The engineers have been one of the most conservative groups in the nation; surely it is no accident that Herbert Hoover has been their foremost spokesman. None of Veblen's critics has provided an adequate

Edwin T. Layton, "Veblen and the Engineers," *American Quarterly*, 14 (Spring 1962), 62–72. Copyright © 1962, Trustees of the University of Pennsylvania. Reprinted without footnotes by permission.

explanation of why Veblen was led to such an improbable belief, though one of them has suggested that the soviet of technicians was no more than an expository device through which Veblen could attack the business order. Just as a fault in the earth's crust enables geologists to gain information about deeper lying strata, so too an examination of the causes of Veblen's error should be of interest not merely in clarifying one of his works, but also because it might provide a new perspective from which to re-evaluate his ideas and methods.

The root of Veblen's interest in engineers is deeply imbedded in one of his most fundamental theoretical assumptions, his notion of "instincts." Instincts, to him, were "innate and persistent propensities of human nature," which along with the material environment, conditioned the habits and conventions which were the very marrow of human institutions. Veblen imagined history as a great dialectic between two instincts: the predatory instinct, or "sportsmanship," and the creative instinct, or "workmanship." Workmanship was characterized by matter-of-fact thinking and behavior; sportsmanship by animism, class distinctions and ceremonial observances. Viewing the contemporary scene, Veblen saw an irrepressible conflict between business and industry. Business represented the predatory instinct; the businessman profited by interrupting or hindering production, that is by "sabotage." Industry represented the creative instinct. The "machine process" itself was a conditioning agency, educating those engaged in productive work in the values and modes of thought of workmanship. As they adopted workmanlike ideas, they rejected the pecuniary thinking of the dominant business culture. "In the nature of the case," Veblen asserted, "the cultural growth dominated by the machine industry is of a skeptical, matter-of-fact complexion, materialistic, unmoral, unpatriotic, undevout." Those conditioned by the machine process, therefore, were the revolutionary group in America.

Prior to 1919 Veblen placed no special emphasis on engineers. He expected modern technology to affect the thinking of the "working class" most markedly, since "they are the most immediately and consistently exposed to the discipline of the machine process." In his earlier works Veblen appeared to have only the haziest notion of who engineers were and what they did. He referred to engineers casually as one of the elements of the working class, lumping them with laborers and mechanics. However by 1919 Veblen's emphasis had shifted to engineers; other groups were relegated to a subsidiary role. "The industrial dictatorship of the captain of finance," Veblen wrote, "is now held on sufferance of the engineers and is liable at any time to be discontinued at their discretion, as a matter of convenience." Moreover, by this time Veblen's treatment of engineers had become more precise. He examined the structure of the engineering profession, distinguishing among consulting engineers, efficiency engineers, production engineers, and even between old and young engineers. And not content with analyzing the function of each type of engineer, Veblen also summarized their thinking. "Right lately," Veblen noted, "these technologists have begun to become uneasily 'class conscious' and to reflect that they together constitute the indispensable General Staff of the industrial system. Their class consciousness has taken the immediate form of a growing sense of waste and confusion in the management of industry.... So the engineers are beginning to draw together and ask themselves, 'What about it?'"

The evolution of Veblen's thought concerning engineers owed less to theory than to actual occurrences. Veblen was aware of and deeply influenced by a chain of events taking place from 1915 to 1920 within one particular engineering society, the American Society of Mechanical Engineers or A.S.M.E. In particular he was impressed by the activities of two mechanical engineers, Morris L. Cooke and Henry L. Gantt. The thoughts and deeds of these two engineers plus certain happenings in the A.S.M.E. provided the empirical foundation for *The Engineers and the Price System.*

Veblen's characterization of the engineers' thinking was an accurate description as far as mechanical engineers were concerned. As Veblen suggested, engineers were becoming class-conscious. *Mechanical Engineering* was full of articles indicating a growing awareness of engineers as a separate group with common ideals, interests and ambitions. Nor were these writers modest; they portrayed the engineer as the mainspring of progress, as the shaper of a new civilization, as something like a general staff. They displayed a boundless confidence in the engineering method, which as one engineer predicted, would carry society "far into the promised land of economic efficiency and social justice." Some engineers were concerned with waste and mismanagement in industry, and were beginning to plan an investigation of the nature and extent of such waste, to be conducted by the engineering profession. Veblen was also correct in asserting that the engineers were drawing together and asking "What About It?" Just two years before Veblen wrote, the leader of a dissident group in the A.S.M.E., Morris L. Cooke, published a pamphlet, *How About It?* and it was subtitled, *Comment on the 'Absentee Management' of the American Society of Mechanical Engineers and the Virtual Control Exerted Over the Society by Big Business—Notably by the Private Utility Interests.*

Morris L. Cooke was the most active leader of a group of young rebels who were questioning the status quo both in industry and the engineering profession. Cooke was enraptured by the idea of applying science to society—that is, what is usually termed "planning." He was convinced that the engineering method might be fruitfully used not only in industry, but in education and government as well, with enormous gains in efficiency. He got a chance to try out some of his ideas when he served as director of public works for the city of Philadelphia from 1911 to 1915. Cooke's efforts met determined opposition from the various utilities serving that city, which were loath to see profitable arrangements upset in the name of efficiency. In particular, Cooke got into a battle with the Philadelphia Electric Company over electric rates. Cooke was embittered on discovering that while the utilities were able to command the services of some of the most eminent members of the engineering profession, the city was virtually unable to obtain the assistance of competent and unbiased engineers. In 1915 Cooke was elected vice-president of the A.S.M.E. and thus became a member of the society's ruling council. There Cooke had ample opportunity to observe something that he already suspected, namely, that the A.S.M.E. was dominated by big business, and especially by the utilities. Cooke was convinced that the low status of the engineering profession was due to this domination by selfish business interests.

Cooke, always the man of action, tried to end the engineers' subordination. As early as 1908 he proposed that engineers should serve the public with the same loyalty that they had previously given to their employers. In 1915 Cooke presented a paper before the A.S.M.E. in which he criticized, in general terms, the pro-utility bias of the engineering profession, pointing out the harmful effect this had on the engineer's status. Engineers affiliated with the utilities attempted to prevent the reading of his paper, and, failing in this, they severely criticized it when presented. A former president of the A.S.M.E. Alexander C. Humphries, suggested that Cooke was unfit to be a member of a profession. Cooke supported his charges in a series of public lectures, which were printed as a pamphlet under the title, *Snapping Cords.* Previously Cooke had tactfully avoided personalities; now he attacked by name several of the most prominent consulting engineers engaged in the utility field, citing in each case specific examples of business bias. His critic Humphries was among those castigated; another was Dugald C. Jackson who had defended the Philadelphia Electric Company in its dispute with Cooke. The engineers attacked by Cooke were able to get him censured by the A.S.M.E. for allegedly unprofessional conduct. This, along with the success of the utility interests in stifling discussion of public policy questions in the A.S.M.E. led Cooke to issue his denunciation, *How About It?*

By 1917 Cooke had emerged as the leader of a faction bent on the overthrow of the pro-business oligarchy controlling the A.S.M.E. Through *How About It?* and other writings he rallied support from the rank-and-file members. In the period 1917–20 Cooke and his supporters achieved a large measure of success; the A.S.M.E. was virtually revolutionized. The rebels carried measures to democratize the society; they severed ties with business organizations; they adopted a new code of ethics which abandoned the old doctrine that the engineer's first professional obligation was loyalty to his employer; they took steps looking toward the unification of the entire engineering profession and its participation in politics. In 1919 a friend of Cooke's, Fred J. Miller, was elected president of the society. Thus by 1919 the rebels virtually controlled the A.S.M.E.

The aim of the insurgents was to engineer society; the reorganization of the engineering profession was no more than a means to that end. But the engineering of society was a sharp break with prevailing ideas, and a new social philosophy was clearly necessary. This new theoretical framework was provided by another mechanical engineer, Henry L. Gantt. Even more than his friend Cooke, Gantt was convinced of the inefficiency and incompetence of the "men of commercial instincts and training" who controlled American business. He estimated that the economic system was operating at only 25 percent efficiency. During 1916, while groping his way toward a more satisfactory social philosophy, Gantt stumbled upon two writers who influenced him; one was Charles Ferguson, an eccentric who believed that experts should rule society, and the other was Thorstein Veblen.

Gantt presented his new philosophy first as articles in the technical press, and then published them as a book in 1919. In these writings Gantt drew a distinction between profit and service analogous to Veblen's business and industry. "Production and not money," Gantt insisted, "must be the aim of our economic system." To achieve this end Gantt proposed that engineers be placed in control. He suggested that they could best operate the economy through public service corporations similar to those employed by the Federal Government during the First World War. Gantt himself had devised a system of charts which, he thought, would greatly facilitate centralized national planning.

Gantt lacked a practical plan for putting his philosophy into practice. In December, 1916 he presented some of his ideas before a meeting of the A.S.M.E. and afterward held an informal gathering attended by 34 interested mechanical engineers. These men formed an organization, the New Machine, electing Gantt president and Charles Ferguson executive secretary. The aim of this group was the "acquirement of political as well as economic power." But the furthest the New Machine got in this direction was to send a letter to President Wilson urging that the control of industry be taken from "idlers and wastrels" and handed over to "those who understand its operations." In 1919 Gantt died.

Veblen probably first became aware of the activities of Gantt and Cooke through a friend at Stanford, a professor of mechanical engineering, Guido Marx. Cooke sent copies of his more significant papers to Marx, and it is likely that he provided Marx with copies of Gantt's works as well. Marx and Veblen were also in the habit of exchanging papers. At any rate Veblen in 1919 became obsessed with engineers.

Veblen's interest in engineers is easily understood. The young rebels had virtually taken over the A.S.M.E. Gantt's New Machine was in effect a "soviet" of technicians, but without a positive program, and, after Gantt's death, without a leader as well. Veblen was aware that Gantt had proposed a study of waste in industry. When teaching at the New School for Social Research in New York, Veblen, through his disciple Leon Ardzrooni, contacted some of the members of the New Machine and suggested Marx as a leader to replace Gantt. Ardzrooni gained the impression that these engineers would be willing to meet with Marx once a week as part of a course at the New School. Veblen persuaded Marx to come to New York to guide the engineers. Marx modestly suggested that Cooke would be the logical man to replace Gantt, but Veblen and Ardzrooni preferred Marx, doubtlessly because he was already something of a convert to Veblenism.

Veblen's attempt to influence the engineers was not successful. This was certainly not due to any lack of energy on the part of Marx. Marx contacted Cooke and arranged for him to give one of the lectures for the course at the New School. He also obtained from Cooke a list of engineers who might be interested in the course and arranged a meeting between Veblen and Cooke. A meeting between the insurgent mechanical engineers and Veblen's group was held. But the engineers were unwilling to accept the leadership of Veblen and Marx. Cooke, though friendly, regarded them as spokesmen for the "extreme left." A group did gather at the New School, and they formed the "Technical Alliance," the ancestor of the technocracy movement of the 1930s; but they were without influence in the engineering profession.

The one lasting result of Veblen's contact with the engineers was his book, *The Engineers and the Price System*. This work is a monument to Veblen's misunderstanding of the events taking place within the engineering profession. Veblen viewed the engineers through the spectacles of his instinct psychology. He assumed that they were being led to reject business culture by the conditioning of the machine process; because they personified the instinct of workmanship they would constitute the spearhead of revolution.

However, the engineers' revolt betokened less a rejection of the traditional culture than an affirmation of it. The engineers' aim was to preserve their middle-class status. Between 1880 and 1920 the engineering profession grew from 7,000 to 136,000, an almost twentyfold increase. But the enlargement of the engineering profession was accompanied by a threat to its status. Most of the new engineers were employees. The consulting engineer, an independent professional man, seemed to be on the road to extinction. Engineers hoped that through reform they might regain their old status. Only through public service, Cooke maintained, could the engineer pull himself out of the "hired servant" class. The insurgent engineers were fighting commercialism, but they were doing so in the name of professional ideals. Professionalism, to the engineers, involved notions of rigid hierarchy and elaborate ceremonialism, and thus fell within the "predatory" culture in Veblen's typology.

Veblen was aware that the engineers had certain conservative tendencies, since he indicated that the vested interests need not fear a revolutionary overturn "just yet." To Veblen these tendencies were merely holdovers from previous modes of thought, indicating that the conditioning process of the machine was not yet complete. But the engineers' conservative ideas were the fundamentals upon which they based their thought. Their aim was to preserve traditional values, such as private property, individualism and Christian morality. The wished to save the existing society, not destroy it; to avert a revolution, not to start one. The engineering of society was no more than a means to this end. Like good designers they wished to use the engineering method to achieve certain "given" objectives with greatest efficiency. Veblen's theoretical bias led him to confuse means with ends. In so doing Veblen imputed to engineers a revolutionary potential which they did not possess.

Another source of Veblen's misunderstanding of engineers lay in his casual, unsystematic methods. Veblen's research does not seem to have gone far beyond the reading of the more prominent works of Gantt and Cooke. Ironically, Veblen was probably deceived by echoes of his own ideas and phraseology in the writings of these two engineers, such as Gantt's distinction between profit and service. Had Veblen read much further he might have arrived at different conclusions. For example, Veblen's analysis of the structure of the engineering profession bore little resemblance to reality. He asserted that it was the production engineers who were leading the revolt against pecuniary ideas, while the consulting engineers and efficiency engineers were the allies of the business leaders. In fact the exact reverse was the case. Almost all of the insurgent engineers were consulting engineers. In the A.S.M.E. the most active leaders, including Cooke and Gantt, were disciples of Frederick W. Taylor, the founder of scientific management. In short the rebels were consultants and efficiency engineers. Veblen's characterization of consultants held for certain restricted cases, among which the most notable, perhaps, was the field of public utility valuation. The consultants in this area were allied with the utility interests; this was the purport of Cooke's complaint in *Snapping Cords*. For the profession as a whole, however, engineers holding managerial and supervisory positions in industry constituted the most important source of business influence. They might well have been characterized as "production engineers."

Veblen called the engineers the "indispensable General Staff" of the industrial system. For some time engineers had been portraying themselves in similarly exalted terms. However, the engineers' statements were less literal descriptions than reflections of an emerging class consciousness. Far from indicating the power of engineers, these served as compensation for the insecurity and powerlessness which engineers actually felt. In fact, the engineers were relatively junior members of the industrial bureaucracy. Nor were they especially indispensable, at least for such short-run considerations as a revolutionary overturn. Veblen saw the engineer as a sort of superworker. In practice engineers typically were engaged in "staff" functions, such as research, design and planning, rather than the "line" activities directly related to production.

Veblen apparently based his generalizations concerning engineers almost wholly on the American Society of Mechanical Engineers. There were reformers active in other major engineering societies, but the essentially conservative character of the insurgent movement was much clearer outside the A.S.M.E. Cooke's counterpart in the American Institute of Mining Engineers, for example, was Herbert Hoover. Indeed by 1919 Hoover had emerged as the national spokesman for all the insurgent engineers. Cooke was a fervent admirer of Hoover, whom he called the "engineering method personified." Shortly before meeting Veblen, Cooke endorsed Hoover for the Republican presidential nomination. "This opportunity to put a rugged, red-blooded, warm-hearted, commonsense, liberty-loving American in the White House," wrote Cooke, "seems well-nigh providential."

Veblen's misunderstanding of the engineers stemmed ultimately from his own position as a transitional figure in the development of social science. In his institutional approach Veblen made a sharp break with the older formalistic school of economics. But Veblen paid a penalty for taking the lead. His break with the past was incomplete. His supposed scientific objectivity provided only the thinnest of veneers for what was essentially a moral critique of the existing order. According to Veblen's instincts, workmanship stood for what was good in human nature and sportsmanship for what was bad. By means of these concepts Veblen introduced into his own theory the same sort of "animistic" thinking that he condemned in the received economics. Veblen's break with the past was incomplete in yet another sense. The institutional approach implied a new methodology— empirical, quantitative, controlled. Such techniques were adopted by Veblen's students, among others, but Veblen himself remained a member of the old speculative, "armchair" school of social scientists. Veblen, therefore, was like the insurgent engineers—the radical propensities of both served to conceal an underlying conservatism.

Human Values and Modern Technology

HERBERT J. MULLER

Man has conquered nature, but he has yet to conquer himself. Modern technology, for all its power, does not adequately serve human values. Indeed, the failure to control technology and direct it to worthwhile ends may be recorded by future historians as the greatest failure of our civilization. In the following selection, Herbert J. Muller attempts to evaluate the gains and losses of industrialism. While freely acknowledging the benefits conferred by modern technology, he understandably stresses the negative. His catalog of horrors could be indefinitely expanded. Clearly, the control of technology is one of the most pressing contemporary problems. Herbert J. Muller is Distinguished Service Professor of English at Indiana University, and he is author of *Children of Frankenstein, A Primer on Modern Technology and Human Values* (Bloomington, 1970).

My main ideas seem to be commonplace, for as I see it, both the goods and the evils that have come out of modern technology are quite obvious. So my only excuse for rehearsing them is that they don't seem obvious enough to most Americans. In particular, Americans seem too little aware of the common abuses of our technology, the very serious problems it has created. So I should warn you that although I have always been given to a depressing on-the-other-hand style, I propose now to dwell chiefly on the "disagreeable" hand—what we're up against. True, I think writers today

Herbert J. Muller, "Human Values and Modern Technology," *Ingenor* (Winter 1969–70), 11–21. Reprinted by permission of *Ingenor* and of the Indiana University Press for quotations from *Children of Frankenstein, A Primer on Modern Technology and Human Values* (Bloomington, Ind., 1970).

too often talk loosely about crises; it seems that no problem is really respectable until it is called a crisis. But even so, I think what most needs to be emphasized is the gravity of our problems—the very real crises confronting us, such as the state of the black ghettos and the deteriorating cities. At any rate, I'm not optimistic about the prospects of our making an adequate national effort to deal with these problems. My point, or my excuse, is that we can no longer afford any easy optimism, and can hope for the best only if we take a hard look at the worst.

Now, my primary concern is human values—values of an old-fashioned kind that at once involve me in commonplaces. [It's significant, incidentally, that values these days are regularly called "human," which is strictly redundant, since only human beings can have conscious values.] Anything that people like or want can be considered a human value. Today no values are more plainly human than the money values that Americans are devoted to. Still, money is plainly only a means to some end. And so is our whole marvelous technology. The matrix of our problems is the common assumption, in effect, that our technology is an end in itself—an assumption fortified by the immense energy that goes into it, the worship of efficiency as the sovereign ideal, the boasts about our material wealth and power, the national goal of steady economic growth, and the national idol GNP (Gross National Product), which I gather now runs to over $800 billion a year. Of this Robert Kennedy remarked, a few months befor his assassination, "The GNP takes into account neither our wit nor our courage, neither our wisdom nor our learning, neither our compassion nor our duty to the country. It measures everything, in short, except that which makes life worthwhile." But this only forces the basic question: What, then, is the proper end for man?

As an old-fashioned type, I'm assuming that the proper end is "the good life." Most Americans, I suppose, would define it as happiness, though with not too clear an idea of what they mean by the term. To me, their pursuit of happiness too often looks like a compulsive, joyless effort to escape boredom. But anyway, a people blessed with far more advantages than any other society has ever enjoyed is not clearly the happiest people on earth. And I am arguing that one reason is their paltry conception of the good life, or what I have called the highest standard of low living in all history. Since most of us are conscientious relativists these days, knowing that no absolute standards can be conclusively demonstrated, I would stress that the good life is nevertheless rooted in absolute goods—good for their own sake, which can't be strictly proved, but which don't need to be proved once they are known. They include such elementary goods as health, physical well-being, and comradeliness, but also such so-called higher goods as the satisfaction of natural curiosity, of the aesthetic sense, and of the related craftsman or creative impulse. These are the source of the traditional values of culture—the pursuit of truth, beauty, and goodness that constitutes the peculiar dignity and worth of man. And in the free societies they have come to include the values of personal freedom and dignity, self-realization, having a mind and a life of one's own.

In this view, it would seem clear that the effects of our technology have been thoroughly mixed, only nothing seems harder to keep clearly, steadily in mind. Literary people tend to forget the obvious goods that have come out of it. Behavioral scientists and technicians tend to slight the no less obvious neglect of fundamental human values.

So let us take a look at the beginnings of modern technology for the sake of some perspective—an historical perspective that I think is too often lacking in both the critics and the champions of our technology. Specifically, let's consider the invention of the railroad, one of the key developments in the Industrial Revolution that had got under way in the late 1700s some 50 years before. In 1835, the editor of the British journal *John Bull* was much alarmed by this latest invention. "Railroads, if they succeed," he warned, "will give an unnatural impetus to society, destroy all the relations that exist between man and man, overthrow all mercantile regulations, and create, at the peril of life, all sorts of confusion and distress."

Well, railroads of course did succeed. By the end of the century they were carrying well over a billion passengers a year, and the editor had long since come to seem like an old fogey. So it's important to see that he was quite right. It's now hard to realize that before the railway era, the great majority of people the world over spent their entire lives in the region where they were born, never leaving the village except to go to the nearest market town, always remaining set in their traditional ways. The railroad accordingly did represent an unnatural impetus, and a profoundly disruptive one. It signalled the end of the old social and political order as it broke down both geographical and social barriers by carrying ever more millions of passengers from all classes. It became the popular symbol of the technological revolution that was in fact creating all sorts of confusion and distress.

More to the point, the editor of *John Bull* spoke for not only many respectable people of his day but for the vast majority of mankind throughout history. Men had always tended to resist fundamental change, any radical innovation. So the railroad symbolized the deepest change that was taking place, which was the attitude toward change itself. It was the growing disposition to accept innovation, even to welcome it. Change was now being called progress. But then we must add that that stodgy editor is still not really a stranger in our revolutionary world. Although change has long since become routine, most people welcome only superficial novelty—the latest models, gadgets, thrills. They still resent and resist any call for fundamental change in their ways of thinking. Or, as Bertrand Russell once observed, "Most people would sooner die than think, and in fact do so." Especially in America, they do not at all welcome radical new ideas. In spite of our boasts about our greatness, Americans seem more afraid of radicals or revolutionaries in our midst than are any other people in the world. And none are more fearful than the business leaders who keep on revolutionizing the economy, accelerating the drive of our technology—leaders who are the most influential radicals of our day.

This brings up an even deeper paradox. The Industrial Revolution was the work of many inventive, enterprising, daring persons, using their heads. As I read history, it was by no means inevitable, automatic, or predetermined by any iron laws of history. Its early course was foretold by no thinker of the time—including Adam Smith, the most acute analyst in the days when machines were growing up. Yet, by the same token, the Industrial Revolution as a whole was quite unplanned. Its pioneers didn't get together and say, "Well, let's try a society based on the machine for a change." Like the inventor of the railroad, they were unconscious revolutionaries who hardly foresaw, much less intended, the profound changes their innovations would bring throughout the whole society. The Revolution illustrated those "vast, impersonal forces" that we hear so much about. It came to seem impersonal and automatic as one invention called out another, and the impetus given to it by the railroad made it well nigh as irresistible as irreversible. And all along, society was even slower to realize the changes that were coming over it—difficulties that we now call "cultural lag" but that, in spite of our knowingness, are as great as ever because of the terrific drive of our technology. Americans today are still slow to face up to the basic problems it creates.

Well, their best excuse is the obvious goods brought by the new technology—so obvious that it is easy for sophisticates to overlook them. Men have had sound reasons for calling the change "progress." The Industrial Revolution meant much more power over the natural environment, a power men have cherished ever since they started chipping flints hundreds of thousands of years ago. It meant a rise in the standard of living, very slow and spotty at first, but in time unmistakable and historically remarkable. As C. P. Snow reminded the literary world, the Industrial Revolution was the only hope of the poor for a decent living—the poor who had always been the great majority of mankind. Likewise it meant an increasing abundance of material goods—goods it is now easy to disparage as merely material, but which are nonetheless real. It seems necessary to repeat to some literary people that it is quite possible to live the good life in a house with plumbing, central heating, and other comforts and conveniences, that they too evidently enjoy. At least few critics of modern civilization seem to be putting on hair shirts. Or if we say that man cannot live by bread alone (as we should), we should also keep in mind the commonplace that man first of all *needs* bread.

It's also hard to realize now that all through history, down to our century, countless millions of people starved to death, and many millions more had to worry about getting enough bread, as they still do in the non-Western world. A recent UNESCO study of the life of a typical Frenchman of the seventeenth century revealed that if he reached the age of 50 he would have lived through two or three famines, and three or four more near-famines. It is worth repeating that America is the first nation in history in which more people die of eating too much than eating too little.

With all this material progress came some clear gains in human freedom (a value I've specialized in). A higher standard of living meant more effective freedom for ordinary people, more opportunity, a wider range of choice and greater power of choice. In Europe, the Industrial Revolution was a major factor in another boon for the common people—free public education. An agricultural society can get along with a mostly illiterate population, as they all did in the past, but an industrial society can't. So even conservatives, fearful of the masses, were in time persuaded to risk sending poor children to school. And only the wealth provided by the new technology made it possible for countries to educate their entire population. In the democratic industrial countries the upshot has been that the common people have been given more rights, opportunities, privileges, and incentives than they ever enjoyed in the past. It's difficult for you students here to appreciate the literally extraordinary outcome in America that you represent: millions of ordinary Americans going to college—a privilege once reserved for a tiny minority. Millions more have a high school education, also a rarity in the past. This means that most young people today choose their own careers, decide where and how they will live; whereas in the always predominantly agricultural societies of the past, the great majority followed automatically in the footsteps of their fathers, remained peasants who lived out their lives in the village.

But now for the "other hand"—the costs of material progress, which to me seem as obvious. From the beginning, the Industrial Revolution brought some glaring social evils, such as women and children working 12 and 14 hours a day in factory and mine. Economic historians are now fond of saying that the historic novelty was not the evils, but the discovery that they *were* evils and then the resolve to do something about them. The masses of people had always toiled and lived in wretched poverty. I think it's important to say this, but I think it's more important to emphasize that social legislation protecting workers came pretty late, after a great deal of confusion and distress, and that it was always fought by business interests and other respectable people, especially in booming America. Likewise the increasing prosperity aggravated the plight of the many millions of people who remained very poor, and who were always the chief sufferers from the chronic depressions and unemployment that came with the new technology. It emphasized another abiding paradox of this technology—that a rational, efficient organization of practical means that could perform wonders has never got around to satisfying the elementary needs for a decent life of a great many people. Today, most affluent America, infinitely the richest society in all history, still has many millions of poor people, and has only begun a piddling war on poverty.

So too with another early consequence of the Industrial Revolution, the growth of ugly, grimy industrial towns—the foulest environment that man has ever created. Slums in the cities grew bigger and fouler because they were considered a natural, normal condition of industrial progress. And again the wealthiest nation on earth has only begun to clear out some of its worst slums, while the black ghettos keep spreading. The millions who all along were condemned to live in these slums may then remind us that the gains in freedom for ordinary people were offset by new compulsions. Millions of people were also condemned to the routines of factory work, day in and day out going through the same mechanical operations—trivial, uncreative, humanly meaningless. In a real sense slaves to their machines when at work, they also had to live where machines were at home. So they lived in such dreary places as Gary, Indiana—once described as a city inhabited by four blast furnaces and a hundred thousand people.

In general, people have been subjected to the compulsions of the new technology in the interest of efficiency and economy—conceived in terms of money values, not of human costs. I won't go into the too familiar story of how industrialism has worked to standardize and regiment people, mechanize and dehumanize life, and has generated massive pressures against the individual that in another view it had liberated. I would note, however, one consequence of the organizational revolution that came with the rise of industrialism: the growth of corporations, ever bigger business, and then of bigger government, bigger organizations or bureaucracy all through the society, now including big universities too. Since Americans are fearful of bureaucracy in government as a threat to their freedom, they should realize that it's just as much bureaucracy in big business, which is somewhat less concerned about the freedom of ordinary people. The men at the top may still recite from the gospel of economic individualism, but this has little meaning to all the little people in or under the big organizations, now suitably identified by holes in punch cards. [I like the signs carried by the students at Berkeley who were protesting that the University administration was treating them like cards in a computing machine. One sign read: I'm a human being: Don't fold, spindle, or mutilate!]

Well, from here on I shall dwell on the situation in America today. As you all know [and I'm talking to the students] but maybe can't actually realize, the forces of science and technology have lately gathered a terrific momentum, producing an unprecedented rate of change. In the generation since the World War, or in the lifetime of you people, we've experienced more sweeping, startling, radical change than whole civilizations went through in a thousand years. Consider for a moment just some of the historic events you have lived through: the dawn of the atomic age; the conquest of outer space, with rockets to the moon; the revolt and rise of the whole non-Western world (where the great majority of mankind lives); the population explosion, the knowledge explosion, and all the other explosions that make exploding seem to be a normal mode of expansion; the many new technical wonders, from television to the wonder drugs that may make over human personalities; the dawn of automation, with computers to take over the brain work; and so on and so on. God knows what may happen the day after tomorrow. Experts in forecasting have drawn up a list of a hundred significant technical innovations that we can confidently expect by the year 2000, and they have speculated on some of the possible social consequences. But it's impossible to say what life will be like then, or even to be sure that there will be any human life on this planet.

Now, the most obvious problem is the appalling destructive power at our disposal. Nuclear weapons make it possible for man to destroy his civilization, or maybe the whole human race—to do all by himself what it once took God to do with a flood. But this points to the most fundamental paradox. Technological change has been accelerating because it's more consciously, deliberately planned than ever before. The many billions the government has been spending on research and development represent by far the mightiest effort in history to direct change. All about us are the elaborate appearances of planning and control. Never before has a society displayed such apparent foresight and will to take charge of its future. Yet the terrific drive seems mechanical and automatic, because it seems irresistible. As is often pointed out (but to no effect), it appears that we *must* do whatever is technologically feasible, no matter what the human cost. So Americans *must* spend many billions of dollars to keep producing bigger and better nuclear weapons, even though we already have many more than enough to blast the whole earth—the equivalent of 14 TNT tons of explosives for every human being on earth. Likewise we *must* spend more billions to put a man on the moon, while a billion or so men on earth don't get enough to eat. Now it seems that we *must* build supersonic planes, no matter what the dangers of sonic boom. They might remind us of the test pilot who radioed in: I'm lost, but I'm making record time.

So the final question is: *Can* man control his terrific technology? Can he direct his immense power to obviously sensible, humane ends—first of all to making the world a safe place to live, then to making it a more civilized place? And since there is so much deliberate planning, at least the appearance of control, the immediate question is: Who is directing our terrific technology, and for what purpose?

Robert Hutchins, a voice crying in the wilderness of California, has given a blunt answer: Our technology is being directed piecemeal by the wrong people, in the wrong ways, to the wrong ends. Specifically, it is being directed in part by our government, primarily for military purposes. For the rest, it is being directed by big business in the interests of private profit, commonly with little regard for any values except money values. To be sure, government is still providing many valuable services, business is producing an abundance of real goods. But again I judge that we need to emphasize much more the glaring abuses of our technology, which again don't seem glaring to most Americans. Our immense power is increasingly a power over not only nature, but over human beings. And for civilized purposes—social, moral, cultural, spiritual—it seems to me a basically irresponsible power.

First, our government. The reason we've spent more than a trillion dollars on military defense including many billions on obsolescent weapons systems such as the ABM's coming up now, is of course, the Cold War—the felt need of keeping ahead of the Russians. The need has also been obvious, in that the Soviet doesn't always act like a peace-loving nation. But even so, it illustrates the appalling folly of collective man. Of course the nuclear arms race gives no real security to either country. Nobody could win a big nuclear war. There is also the cold truth that in our obsession with the Cold War, we devote to our Arms Control and Disarmament Agency only a tiny fraction of one percent of what we spend on military hardware—and that, to the rest of the world we don't always look like a peace-loving nation either.

For the purposes of a free society, still another troublesome consequence is the notorious military industrial complex that President Eisenhower warned the nation about. Life and death decisions affecting all of us are made by a very few men now often in secrecy without public debate. As Aldous Huxley wrote, "Never have so many been so completely at the mercy of so few." As for our war in Viet Nam (one of those decisions made in secrecy), I happen to be among those who consider it a frightful mistake, but in any case the war came first. Among its casualties were the public welfare programs, the fate of which makes it indecent to talk anymore about a "Great Society." The cry in Washington is still that we can't afford a serious war on poverty or the slums.

This brings up another elementary neglect of human needs that Galbraith dwelt on in *The Affluent Society*. As we grew richer by the year, all the time boasting of our efficiency, our postal service grew poorer, our schools and hospitals more overcrowded, our streets dirtier, our rivers and the air we breathe more polluted, our cities more unfit to live in. Congressmen who didn't bat an eye over appropriating $60 billion for the military always attacked as "reckless extravagance" proposals to spend a few billions on public welfare programs—whereas we were in fact spending a smaller proportion of our growing national income on public services than we had been and the supposedly alarming increase in the Federal debt was actually a decrease in relation to this income. In general, the sloganeering that passes for thinking in the nation's capital is based on assumptions about our economy and government that were not very realistic even in the past, but that an affluent society has made hopelessly obsolete. Hence, while serious thinkers have grown much more alert to rising problems than men were in the early days of the Industrial Revolution, government remains slow in responding to them. The latest example is the riots of the blacks in the city ghettos.

To come closer to home, let's consider another revolutionary consequence of the affluent society: the education explosion that has sent hordes of students to college. Now that the government is pouring billions into the universities for research and development, educators are talking of another major industry—the knowledge industry—which has become the biggest industry in the country. In the kind of technical language that has grown fashionable, Clark Kerr writes that "the production, distribution, and consumption of knowledge" now accounts for 29% of the Gross National Product, and is growing at about twice the rate of the economy. Daniel Bell adds that the university bids fair to be the primary institution of our new kind of technocratic society, even more important than the corporation. All of us here might therefore rejoice in our new eminence. All, that is, except those of us engaged in the old-fashioned business of a liberal education. The consumption of our kind of product represents a small fraction of the GNP. For the emphasis is on specialized technical knowledge, not on breadth of interest or understanding, enlargement of mind, enrichment of personality, or the cultivation of civilized values. The billions are going chiefly, of course, into science and technology. When President Johnson requested a few million dollars for the new Foundation on the Humanities and the Fine Arts, Congress promptly slashed this tiny amount in half. A Republican leader who tried to cut it out entirely protested, "We got along pretty well in this country for a century or so without spending money for culture and the humanities." In short, culture is a mere frill.

As for the serious business of life, all that is supposed to keep America healthy and strong, the leaders of both parties appear to agree that our primary national goal must be steady economic growth. This might seem odd, since the basic economic needs of most white Americans have long since been satisfied by our technology, and business can flourish only by creating a lot of superfluous needs, such as the latest model and the latest gadget. But I suppose business must come first in the new kind of society we've developed, in which we are all more dependent on the state of the economy than ever before. I'm reminded of the prayer of the little English boy in the World War: "God bless Mother and Daddy, my brother and sister, and save the King—and oh God, do take care of yourself, because if anything happens to you we're all sunk." And so *we're* all sunk if our economy collapses, and we may be in a bad way if it grinds to a halt—still worse if it enters an old-fashioned depression of the kind it had regularly in the palmy days of free private enterprise.

But even so, I stick to my simple point, that steady economic growth is only a means to some end. Since business seems to be taking pretty good care of itself, it's all the more important to have some civilized notion of the end or of the good life. Now we'll never agree, of course, on just what is the good life for man, any more than the wise or the holy men through the ages have ever agreed. And I, for one, don't want complete agreement, just because man has such rich diverse capacities, and because I'm committed to the values of individuality, or people thinking independently for themselves. As a teacher of the humanities, I no doubt share their common objective—how to be more like me—but I wouldn't even press this laudable objective. Still, thoughtful people can at least agree that wealth and power don't automatically lead to a better life. And since I began by saying that we all know some absolute goods, I will now add that we all can and do agree on some absolute evils.

An obvious example is the pollution of our rivers and our air. Nobody would deny that this is a bad thing. Then the pertinent question is: Why has affluent America put up with it for so long and only just begun to do something about it? And again the answer is obvious. Our great industries produced the pollution. To reduce it will require government interference with private enterprise, cost a lot of money, and cause businessmen some trouble. In other words, it calls for more "creeping socialism"—what appears to give businessmen nightmares. I must confess that I find it hard to share these nightmares. Inasmuch as they've gone on piling up record profits, corporations have grown into giants, and my impression remains that General Motors is not a poor, helpless, grievously oppressed organization. I state as an historical fact that never in any past society did businessmen as a class enjoy such power, privilege, and opportunity as they do in America today. Then I'd add that they have more responsibility too.

So I'm brought to the role of big business in directing our wondrous technology. In the old days before the New Deal (the golden age of Barry Goldwater), it was often frankly irresponsible. J. P. Morgan said in so many words: "I owe the public nothing." Another big magnate said more bluntly, "The public be damned!" Now I think the big corporations have grown distinctly more responsible. But I also think that too often they're by no means responsible enough in the exercise of their very great power. Their spokesmen still tell us that economic freedom is the most fundamental kind, the heart of the American way of life, even though the authors of the Constitution neglected to mention it. But what they defend most zealously is the freedom to be socially irresponsible.

Here I won't go into some of my own pet irritations, such as all the commercials the big TV networks now work into their news programs; or the scandalous income tax favors lavished on the highly protected oil industry, which repays us with Texas millionaires [types who may recall the old saying that self-made men relieve God of an awful responsibility]; or the scandalous profits of the drug industry [a matter I feel bitter about because of a blood pressure pill my doctor has prescribed—a drug that the government bought from a European country for the equivalent of five cents a hundred pills, and that costs me $8.75 a hundred in the drugstore]. At the risk of being indiscreet or un-American, I'll consider just your mammoth, automobile industry. The automobile, which has been doing away with the passenger train, has brought much more of the "peril to life" that the editor of *John Bull* feared in the railroad; many more Americans have been killed on the highways than in both world wars. Now, under government pressure, the industry has only begun grudgingly to put more safety devices into its cars, and it is still resisting efforts to make it add exhaust controls in order to reduce pollution of the air. Why this indifference to human safety and public health? Alfred Sloan, the wizard who built up General Motors, stated the basic reason quite simply in his memoirs: "The primary object of the corporation was to make money, not just to make motor cars." It made more money by making expensive cars with excess power and chrome, annual new models with only styling changes, instead of just sturdy, efficient cars.

Well, profits must of course be the primary concern of big corporations. First and last they're all necessarily out to make money. Andrew Hacker, a kindly disposed political scientist, has explained that they can't be held socially responsible for their decisions about how to keep their business profitable, and that social problems created by their technological advances, such as automation, must be left to the government. But then the trouble is that their spokesmen habitually resist government efforts to deal with the problems, just as they have fought every major reform in this century. Like the automobile manufacturers, they protest that the government is interfering with business. Likewise they too have opposed public welfare programs as reckless extravagance, rehearsing the slogans about spending ourselves into bankruptcy while taxing ourselves to death.

Meanwhile, business has built up another revolutionary enterprise: a $15 billion advertising industry to make Americans ever more extravagant in satisfying their private or selfish wants. It is dedicated to the proposition that selling eyewash in fancy packages is what keeps America healthy and strong. This advertising industry brings us back to the curious tyranny of our kind of economy. It is absolutely essential, I suppose, to the steady economic growth on which we now depend. Americans *have* to be sold on the latest model and all kinds of fancy or superfluous goods, or else our economy will collapse. As has often been said, the primary function of Americans today is to be consumers. They must keep on consuming faithfully, arduously, to the end of their days—an end suitably marked by a costly funeral. Theirs not to question why, theirs but to go and buy. Then it must be said that Americans appear to be doing their duty gladly enough.

So we come to the root of the problem. We cannot simply blame either government or business for the abuses of our technology and the neglect of fundamental human values. The fault is finally the fault of the American people. They have not been protesting against the folly of the nuclear arms race. They are not demanding that we spend anything like the billions it would take to make the American city fit to live in, or to provide decent homes, schools, and jobs for the blacks in its ghettos. Still less do they call for more aid to the billion people on earth who don't get enough to eat. [Let me add that I'm sparing you the biggest problem looming up: the growing gap between the few "have" nations and the many "have-nots," made worse by the population explosion. So just the annual increase in our national income has been greater than the total income of the whole continent of Africa, with its more than 300 million people.] Americans are plainly in a conservative mood, disposed to sit tight because they're sitting pretty, and to complain only about the taxes they have to pay for having it so good. They are in no mood to make any real sacrifices, even though their taxes amount to a smaller proportion of the national income than the taxes levied in the European democracies. And in their self-indulgence, the great majority approve the values they're sold in the interest of private profit. Businessmen can always defend the cars with excess power and chrome and fins, trashy programs on TV, and all the other trivial or tawdry products of our technology, by saying that they are only giving the people what they want. Such living values make uglier the violence of the white backlash against the blacks, the uses of white power to keep the blacks out of the suburbs. What are these people trying to preserve? What vision? What ideal? What image of freedom and dignity? What notion of the good life?

Let us look at Americans in their automobiles—the sacred cows of the American way of life. Another reason for the blight of the American city is that its government has been spending most freely to take care of the needs of the automobile, eating out the heart of the city to provide more parking space while building more throughways leading into it, only to increase the congestion. To me there is no more absurd spectacle than that enacted every morning and afternoon in America—millions of people driving to and from work in the city, one to a car, bumper to bumper, often through smog, often tense and irritable, unable to relax and enjoy the passing landscape. Worse, they seem unaware that there's usually little to enjoy in this bulldozed landscape anyway, chiefly a clutter of billboards and neon signs, garish shopping centers, drab or ugly suburban sprawl, and honkytonk and other commercial slops to make more hideous the approaches to our cities. Apparently Americans accept ugliness as natural and normal, just as people did in the bad old early days of the Industrial Revolution. They don't mind the transformation of God's own country into "God's own junkyard."

So I'm brought back to the basic question: Can we control our terrific technology? Can we direct it to humane, civilized purposes, a more gracious way of life, more pleasing to eye and ear? I think it is technically possible to do so, in view of all the deliberate direction. I also doubt that there will be anything like a great national effort to do so, certainly not while we're obsessed with the Cold War, but most likely not even if we're ever at peace with the world. Robert Hutchins concluded that to direct our technology intelligently, first of all to get it out of the hands of the wrong people, would call for a revolution, not only economic and political, but moral and intellectual. I see no immediate prospect of such a revolution, or any popular demand for one in America. In popular journals there has been quite a fashion lately of forecasting what America will be like in the year 2000, and the basic picture is always the same: the future will bring only a bigger and fancier technology, only a wonderland still more fit for Alice. Its capital has been given the appropriately ugly name of Boswash, a huge metropolitan sprawl or megalopolis reaching all the way from Boston to Washington.

In a sense, I'm more pessimistic than Hutchins. His charge that the wrong people are in charge of our technology forces another question: Who then are the *right* people to put in charge? I see no possibility of agreement on an answer to this question, nor do I myself have a clear answer. I would never put our government, *any* government, in complete charge of our technology, any more than I would turn it over completely to private business. Neither would I entrust the job to any central organization, any body of social engineers or systems analysts, for these specialists always tend to put efficiency or the needs of the system above the human values I'm most concerned about. And I would not turn over the job even to a foundation of the humanities, headed by Robert Hutchins. From long experience as a professor, I can only imagine the endless wrangle over the right means and ends, the noisy confusion that led an African observer of an American faculty meeting to ask, "Do you think they're really ready yet for self-government?"

So again I conclude lamely, as I did the last time out. So long as we remain anything like a free society, we'll go on directing our technology piecemeal, at best curbing some abuses, making somewhat ampler provisions for both elementary human needs and civilized human values. Finally, our best hope remains the obvious one: education. All of us here know that education is a very long, slow, uncertain process. In the universities it's complicated by the increasing specialization which is turning out more one-dimensional men. This might make us all wonder whether we're preparing students for life the day before yesterday, instead of tomorrow—and then wonder just how to give them the professional training they need and also give them a better understanding of our technological society, make them good citizens with a proper concern for human values, help them to make satisfying use of the increasing abundance and leisure, develop them as independent individuals with minds of their own, and prepare them for responsible leadership in a future that the specialists in serious forecasting tell us will be still fuller of problems.

As an educator too I can't end on a high note. The best I can do is to point to some hopeful signs. The American people seem to be growing more alert to the changes in our society. They've been making best sellers of books about the organization man, the hidden persuaders, the status seekers, the lonely crowd, the pressures to conformism—the not-so-brave new world. Of course, there is much more to be said on behalf of the American people. In the universities there is a considerable stir, more of an awareness of what H. G. Wells said long ago, after World War I: "From now on, history is a race between education and catastrophe."

Yet at the end I'd repeat that there will have to be a great deal more change if our world is to be made safe and fit for human beings to live in, and that our pace of adjustment is still pretty slow in view of the accelerating pace of technological change and the urgency of some of our problems. Then the best I can say is that the wonderful, fearful potentialities we have developed through science and technology are still open, for better and for worse. It is much too easy to blame everything on technology, for what we make of it is strictly up to us. In particular it is up to the younger generation, who will inherit both the powers and the messes created by their elders. So among the most hopeful signs to me are the signs that more students are growing alert to the problems, taking an active interest in them, aspiring to more than a good, safe berth in a good, safe corporation. Certainly, most of the students I teach are better informed about the state of America and the world than were the college students of my own generation. They include young radicals who consider me just an old liberal (the dirtiest word in their lexicon), but even so I welcome their indignation, their ardor—at least until they begin busting up the campus. In my usual style I might then say: God bless our youth, or maybe God help them—that is, if there be a God, which I don't know either. But I think I know that they may at least still regard the problems as challenges (another of our favorite words today), or even as opportunities. I thank you.

Selective Bibliography

The literature in the history of technology is vast, scattered, and of uneven quality. Fortunately, some good bibliographies are available. There is no better place to begin than with Eugene S. Ferguson, *Bibliography of the History of Technology* (1968). Another outstanding guide, specifically concerned with American technology down to 1850, is Brooke Hindle, *Technology in Early America: Needs and Opportunities for Study* (1966, paperback*). A series of extensively annotated bibliographies on the relations of technology to society has been prepared by the Harvard University Program on Technology and Society and published in its *Research Review*, particularly "Technology and Work," no. 2 (1969); "Technology and Values," no. 3 (1969); Technology and the Polity," no. 4 (1969); "Technology and the City," no. 5 (1970); "Technology and the Individual," no. 6 (1970); and "Technology and Social History," no. 8 (1971). Only the last, however, is explicitly concerned with historical writings. For American agriculture an excellent, annotated bibliography is Carroll W. Pursell, Jr. and Earl M. Rogers, *A Preliminary List of References for the History of Agricultural Science and Technology in the United States* (1966).

There is no satisfactory synthesis of American technology as yet. Perhaps, the best is a work not specifically devoted to American history, Melvin Kranzberg and Carroll W. Pursell, Jr., eds., *Technology and Western Civilization* (2 vols., 1967); the second volume emphasizes American technology since 1900. Roger Burlingame's works are still useful for reference purposes, notably his *March of the Iron Men* (1938), and *Engines of Democracy* (1940). John Oliver, *History of American Technology* (1956) is disappointing. There is useful material on technology in in-

---

*The word "paperback" indicates that the work is now available in paperback. If only the word "paperback" appears after the work cited, its original publication was in paper.

dustry, agriculture, and the military in David D. Van Tassel and Michael G. Hall, eds., *Science and Society in the United States* (paperback).

There are several interpretative studies of particular aspects of American technology. A point of departure is provided by George H. Daniels, John G. Burke, and Edwin T. Layton, Jr., "Symposium: The Historiography of American Technology," *Technology and Culture*, 11 (January 1970). John A. Kouwenhoven, *The Arts in Modern American Civilization* (1948, paperback) originally published under the title *Made in America*, argues that American technology and art were shaped by a democratic, "vernacular" style. Dirk J. Struik, *Yankee Science in the Making* (1948, paperback) is a neo-Marxist interpretation, confined to New England before the Civil War. John B. Rae, "The 'Know-How' Tradition: Technology in American History," *Technology and Culture*, 1 (Spring 1960), stresses the role of business. Edwin T. Layton, Jr., "Mirror Image Twins: The Communities of Science and Technology in 19th-Century America," *Technology and Culture*, 12 (October 1971) describes the interaction of science and technology. Elting E. Morison, *Men, Machines, and Modern Times* (1966, paperback) is concerned with resistance to technological change.

America was a consumer of imported technologies for much of its history. The colonial dawn of American industry is studied in Edward N. Hartley, *Ironworks on the Saugus* (1957, paperback), a work combining traditional history with the new field of industrial archaeology. Norman B. Wilkinson, "Brandywine Borrowings from European Technology," *Technology and Culture*, 1 (Winter 1963) examines one of the most important focal centers of modern industrial development in America. The recruitment of technologists from Britain is explored by Carroll W. Pursell, Jr., "Thomas Digges and William Pearce: An Example of the Transit of Technology," *William and Mary Quarterly*, 21 (October 1964). The same author's *Early Stationary Steam Engines in America: A Study in the Migration of a Technology* (1969) is one of the few books specifically concerned with the transfer of technology.

Before there could be an Industrial Revolution there had to be a revolution in transportation. George Rogers Taylor, *The Transportation Revolution, 1815–1860* (1951, paperback) surveys the revolution in transportation and its implications for industry. Louis C. Hunter, *Steamboats on the Western Rivers: An Economic and Technological History* (1949) is a classic on the steamboat. John H. White, Jr., *American Locomotives: An Engineering History, 1830–1880* (1968) is concerned with the evolution of the railroad locomotive. Forest G. Hill, *Roads, Rails, and Waterways: The Army Engineers and Early Transportation* (1957) deals with the important role of Army engineers in developing modern forms of transportation.

The "American System" of manufacturing was one of America's most important contributions to technology. Two older studies still of value are Joseph W. Roe, *English and American Tool Builders* (1916) and Roger Burlingame, *Backgrounds of Power: The Human Story of Mass Production* (1949). Nathan Rosenberg, ed., *The American System of Manufactures* (1969) and his "Technological Change in the Machine Tool Industry, 1840–1910," *Journal of Economic History*, 23 (December 1963) are modern classics. Much of the interest in mass production has centered about its use in the automobile industry. John B. Rae, ed., *Henry Ford* (1969, paperback) surveys the divergent interpretations of Ford. Rae's *The American Automobile* (1965, paperback) and *American Automobile Manufacturers* (1959) are basic for understanding the automobile industry.

One way to study American technology is to look at particular industries. W. Paul Strassman, *Risk and Technological Innovation: American Manufacturing Methods During the Nineteenth Century* (1959) surveys four important industries: textiles, iron and steel, machine tools, and electrical manufacturing. Two of these are discussed at greater length in special studies: Harold C. Passer, *The Electrical Manufacturers, 1875–1900* (1953) and Peter Temin, *Iron and Steel in Nineteenth-Century America, An Economic Inquiry* (1964). The history of the oil industry is detailed in Harold F. Williamson, and others, *The American Petroleum Industry: The Age of Illumination, 1859–1899* (1959) and *The American Petroleum Industry: The Age of Energy, 1899–1959* (1964). John B. Rae, *Climb to Greatness: The American Aircraft Industry, 1920–1960* (1968) surveys the aircraft industry. Mining is examined in Rodman Wilson Paul, *Mining Frontiers of the Far West, 1848–1880* (1963, paperback).

American farmers were as eager to apply modern technology as were American industrialists. An old, but still useful economic analysis is Leo Rogin, *The Introduction of Farm Machinery in Its Relation to the Productivity of Labor in the Agriculture of the United States During the Nineteenth Century* (1931, paperback). The story of the reaper and its inventor is examined in William T. Hutchinson, *Cyrus Hall McCormick*, 2nd ed. (2 vols., 1969). Reynold M. Wik, *Steam Power on the American Farm* (1953) discusses another aspect of mechanization.

Building and construction was another of the fields where distinctive American traditions of technology appeared. A superb survey, covering all types of building from homes to bridges, is Carl W. Condit's two volumes, *American Building Art: The Nineteenth Century* (1960) and *American Building Art: The Twentieth Century* (1961). A shorter synthesis by the same author is *American Building* (1969, paperback). For the colonial period, Anthony Garvan, *Architecture and Town Planning in Colonial Connecticut* (1951) is outstanding. The origin of the skyscraper is examined by Carl W. Condit, *The Chicago School of Architecture* (1964).

With the steady accumulation of knowledge, technological leadership has passed successively from craftsmen to inventors to engineers. Thus, in addition to looking at particular industries, we can approach the study of American technology by an examination of the occupational and social roles played by technologists. The origins of a tradition of craftmanship in the colonial period is surveyed by Carl Bridenbaugh, *The Colonial Craftsman* (1950, paperback). Esther Forbes, *Paul Revere* (1942, paperback) is a biography of a famous craftsman, who was also an important innovator. The evolution of the technological community from craftsman to inventor and engineer is documented in Eugene S. Ferguson, ed., *Early Engineering Reminiscences, 1815–1840, of George Escol Sellers*, United States National Museum Bulletin, no. 238 (1965).

The inventor as a social type was largely a product of the nineteenth century. L. Sprague De Camp, *The Heroic Age of American Invention* (1961) is a popular account with useful information on the patent system. The difficulties of early inventors are suggested by Greville and Dorothy Bathe, *Oliver Evans: A Chronicle of Early American Engineering* (1935). The evolution of industrial research is surveyed in Howard R. Bartlett, "The Development of Industrial Research in the United States," in U.S. National Resources Committee, *Research—A National Resource: II—Industrial Research* (1941). John Jewkes, David Sawers, and Richard Stillerman, *The Sources of Invention* (1961, paperback) is critical of the common assumption that the lone inventor has been displaced by team workers in laboratories. But no one questions the importance of research laboratories; Kendall Birr, *Pioneering in Industrial Research: The Story of General Electric Research Laboratories* (1957) deals with one of the most significant. Matthew Josephson, *Edison* (1959, paperback) is the biography of one of the founders of team research, who was also America's greatest inventor. The changing role of the inventor in the modern age is examined by two recent and outstanding studies: Thomas P. Hughes, *Elmer Sperry: Inventor and Engineer* (1971) and James E. Brittain, "The Introduction of the Loading Coil: George A. Campbell and Michael I. Pupin," *Technology and Culture*, 11 (January 1970).

The engineer represents the scientifically trained, professional technologist of the modern age. Daniel H. Calhoun, *The American Civil Engineer: Origins and Conflict* (1960) deals with early professionalism and its conflicts with commercial domination of technology. Monte A. Calvert, *The Mechanical Engineer in America, 1830–1910* (1967) is concerned with the role of an elite in the professionalization of mechanical engineering. Raymond H. Merritt, *Engineering in American Society, 1850–1875* (1969) is unreliable but has some useful information on education. Edwin T. Layton, Jr., *The Revolt of the Engineers, Social Responsibility and the American Engineering Profession* (1971) is concerned with the efforts of twentieth-century American engineers to influence the direction of technology.

The studies of professional development indicate that technologists have not been able to control technology. There is much evidence that economic forces provide the primary direction. Jacob Schmookler, *Invention and Economic Growth* (1966) argues that patents follow the market. The role of labor scarcity in promoting labor-saving technology in America has been debated by H. J. Habakkuk, *American and British Technology in the Nineteenth Century: The Search for Labour-Saving Inventions* (1962, paperback) and Peter Temin, "Labor Scarcity and the Problem of American Industrial Efficiency in the 1850's," *Journal of Economic History*, 26 (September 1966). Louis C. Hunter, "Influence of the Market upon Technique in the Iron Industry in Western Pennsylvania to 1860," *Journal of Economic and Business History*, 1 (February 1929) is a classic study.

The impact of technology on American writers is surveyed in Leo Marx, *The Machine in the Garden: Technology and the Pastoral Idea in America* (1964, paperback), which shows the utility of methods of analysis derived from literary criticism. Alan Trachtenberg, *Brooklyn Bridge, Fact and Symbol* (1965) is a multidimensional study of the influence of a great bridge on American consciousness. Thomas Reed West, *Flesh of Steel: Literature and the Machine in American Culture* (1967) deals with more recent literature. A suggestive short study is Hugo A. Meier, "American Technology and the Nineteenth-Century World," *American Quarterly*, 10 (Summer 1958). The social effects of recent technology have been surveyed by Lloyd Morris, *Not So Long Ago* (1949), with emphasis upon the role of motion pictures, the automobile, and radio. A recent study of the automobile's influence is James T. Flink, *America Adopts the Automobile, 1895–1910* (1970).

The group most directly influenced by modern technology has been the workers. An older study of continuing value is Norman Ware, *The Industrial Worker, 1840–1860: The Reaction of American Industrial Society to the Advances of the Industrial Revolution* (1924, paperback). Charles L. Sanford, "The Intellectual Origins and New-Worldliness of American Industry," *Journal of Economic History* 18, no. 1 (1958) examines early efforts to preserve America's moral climate in an industrial age. Elizabeth F. Baker, *Technology and Woman's Work* (1964) deals with the changing role of women in industry. Sociologists and allied scholars have produced a large body of literature dealing with work and workers that contains material of value to students of history. An introduction to this literature is provided by Charles R. Walker, ed., *Technology, Industry, and Man: The Age of Acceleration* (1968). Robert Blauner, *Alienation and Freedom: The Factory Worker and His Industry* (1964, paperback) shows that technologies associated with different historical stages of industrial development have produced varying amounts of worker alienation.

One of the most important forces influencing work and workers has been scientific management. Frank B. Copley, *Frederick W. Taylor, Father of Scientific Management* (2 vols., 1923) is still the standard biography of Taylor. Hugh G. J. Aitken, *Taylorism at the Watertown Arsenal* (1960) is an insightful study of broader scope than its title indicates. Milton J. Nadworny, *Scientific Management and the Unions, 1900–1932* (1955) deals with the development of management thought as well as labor's reaction to Taylorism.

The relation of technology to American politics has only just begun to be studied. Nelson M. Blake, *Water For Cities: A History of the Urban Water Supply Problem in the United States* (1956), is concerned with the politics of water supply during the nineteenth century. Two outstanding studies examine the influence of technology on political thought in the progressive era: Samuel P. Hays, *Conservation and the Gospel of Efficiency: The Progressive Conservation Movement, 1890–1920* (1959, paperback) and Samuel Haber, *Efficiency and Uplift: Scientific Management in the Progressive Era, 1890–1920* (1964). Donald C. Swain, *Federal Conservation Policy, 1921–1933* (1963) continues the story of conservation through the 1920s. Henry Elsner, Jr., *The Technocrats: Prophets of Automation* (1967) is the best available account of Technocracy. Carroll W. Pursell, Jr., "The Farm Chemurgic Council and the United States Department of Agriculture, 1935–1939," *Isis*, 60 (Fall 1969) is a particularly noteworthy study of technology and politics. The relation of technology to war is a subject of increasing interest. Robert V. Bruce, *Lincoln and the Tools of War* (1956) covers the Civil War. James P. Baxter, *Scientists Against Time* (1946, paperback) surveys weapons development during World War II. Harold L. Nieburg, *In the Name of Science* (1966, paperback) is concerned with recent activities of the military-industrial complex.

The influence of American technology abroad is a promising field for study. Merle Curti and Kendall Birr, *Prelude to Point Four: American Technical Missions Overseas, 1838–1938* (1954) is a pioneering work. Marvin Fisher, *Workshops in the Wilderness: The European Response to American Industrialism, 1830–1860* (1967) deals with the reactions of European travelers. Merle Curti, "America at the World Fairs, 1851–1893," *American Historical Review*, 55 (July 1950) examines one of the most important means that Europeans were made aware of American technology. D. L. Burn, "The Genesis of American Engineering Competition, 1850–1870," *Economic History*, 2 (January 1931) shows how American technology influenced Britain. Eugene B. Skolnikoff, *Science, Technology, and American Foreign Policy* (1967, paperback) is a political scientist's account of recent developments.

73 74 75   10 9 8 7 6 5 4 3 2 1